Number Sense

For each picture, tell if the number is used to count, to label, to measure, or to tell order.

1.

_____

2.

_____

3.

24 checkers

_____

4.

Police #707

_____

5.

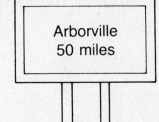

Arborville 50 miles

_____

6.

_____

7.

5th Annual School Olympics

_____

8.

milk 1 Quart

_____

9.

_____

10.

Scoreboard

Home | Visitors
6 | 0

_____

11.

1st prize

_____

12.

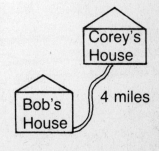

Corey's House

Bob's House

4 miles

_____

Addition and Subtraction

Use the rule given. Find the missing numbers.

1.

	+7
3	__
__	11
5	__
2	__

2.

	−3
__	6
8	__
5	__
__	4

3.

	−5
10	__
6	__
5	__
__	4

4.

	+5
3	__
__	9
2	__
7	__

Look at the problems in Tables 1–4. Which of these can you do mentally?

Color the shape for each missing number.

Missing Number	6	7	8	9	10
Color	Brown	Blue	Purple	Orange	Yellow

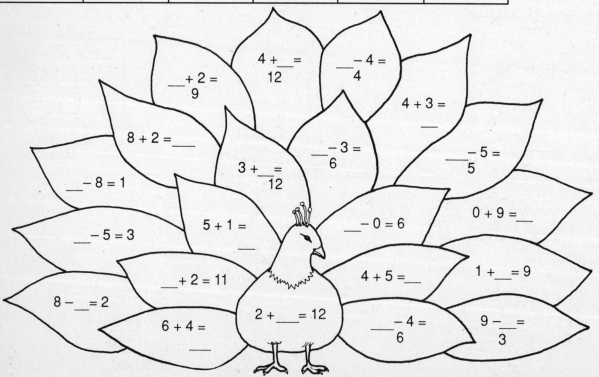

$\_ + 2 = 9$

$4 + \_ = 12$

$\_ - 4 = 4$

$8 + 2 = \_$

$4 + 3 = \_$

$3 + \_ = 12$

$\_ - 3 = 6$

$\_ - 5 = 5$

$\_ - 8 = 1$

$5 + 1 = \_$

$\_ - 0 = 6$

$0 + 9 = \_$

$\_ - 5 = 3$

$\_ + 2 = 11$

$4 + 5 = \_$

$1 + \_ = 9$

$8 - \_ = 2$

$2 + \_ = 12$

$\_ - 4 = 6$

$9 - \_ = 3$

$6 + 4 = \_$

Operation Sense

Write which number or symbol is missing. Then find each answer in the code. Write each code letter above the correct exercise number at the bottom of the page. The letters will solve this riddle:

Four friends heard some thunder.
Four friends tried to get under
One umbrella, so tiny and yet,
Why didn't the friends get wet?

1. $7 - \underline{\quad} = 5$

2. $13 \underline{\quad} 4 = 9$

3. $\underline{\quad} = 5 + 6$

4. $6 \underline{\quad} 4 = 10$

5. $5 + 3 = \underline{\quad}$

6. $7 + \underline{\quad} = 13$

7. $4 \underline{\quad} 4 = 8$

8. $6 - \underline{\quad} = 3$

9. $\underline{\quad} = 13 - 5$

10. $9 \underline{\quad} 4 = 5$

11. $7 + \underline{\quad} = 9$

12. $6 + \underline{\quad} = 10$

13. $5 \underline{\quad} 7 = 12$

14. $\underline{\quad} - 7 = 1$

15. $5 + \underline{\quad} = 9$

16. $\underline{\quad} + 8 = 14$

17. $4 + \underline{\quad} = 8$

Code:	−	+	2	3	4	6	8	11
	D	T	L	R	A	N	I	O

Exercise Number

___ ___   ___ ___ ___   ___ ___ ___
5   4   10 9   2   6   3   13

___ ___ ___ ___   ___ ___   ___ ___ ___.
8   15 14 16   17 7   12 11 1

Using a Problem-Solving Guide

Answer each question about this problem.

There are 12 cows in a field. Of these cows,
5 are Guernseys. How many are not Guernseys?

1. What facts are given?

_____

_____

2. What is the key idea?

_____

_____

3. Rewrite the problem in your own
words.

_____

_____

4. What strategy would you use to
solve the problem?

_____

_____

5. What is your answer to the
problem?

_____

_____

6. Write a way that you can check
your answer.

_____

_____

Solve each problem. Use the
Problem-Solving Guide to help you.

7. On Saturday, there were 14
horses in a field. On Sunday
there were only 8. How many
were taken out?

_____

8. Of the 8 horses left in the field, 3
had white blazes on their faces.
How many did not?

_____

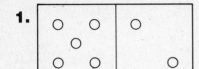

Families of Facts

Write the family of facts for each picture.

1.

2.

3.

_____

_____

_____

_____

_____

_____

_____

_____

_____

_____

Write each answer. Circle the fact that does
not belong to the family.

4. 13 − 8 = _____   **5.** 3 + 4 = _____   **6.** 9 + 6 = _____

   5 + 8 = _____   4 + 3 = _____   15 − 6 = _____

   9 + 8 = _____   7 − 4 = _____   13 − 6 = _____

   8 + 5 = _____   8 − 4 = _____   15 − 9 = _____

   13 − 5 = _____   7 − 3 = _____   6 + 9 = _____

Write the family of facts for the numbers in
each exercise.

7. 9, 3, 6   **8.** 2, 9, 11   **9.** 4, 4, 8   **10.** 14, 6, 8

_____   _____   _____   _____

_____   _____   _____   _____

_____   _____   _____   _____

_____   _____   _____

SHARPEN
YOUR
SKILLS

Even and Odd

Write whether each number is odd or even.

1. 456   **2.** 789   **3.** 23   **4.** 68

_____   _____   _____   _____

5. 53   **6.** 965   **7.** 637   **8.** 12

_____   _____   _____   _____

9. 76   **10.** 531   **11.** 94   **12.** 387

_____   _____   _____   _____

Find each sum. Write whether the sum is odd
or even.

13. $\begin{array}{r} 6 \\ +3 \\ \hline \end{array}$ **14.** $\begin{array}{r} 8 \\ +2 \\ \hline \end{array}$ **15.** $\begin{array}{r} 2 \\ +7 \\ \hline \end{array}$ **16.** $\begin{array}{r} 3 \\ +9 \\ \hline \end{array}$

17. $\begin{array}{r} 1 \\ +7 \\ \hline \end{array}$ **18.** $\begin{array}{r} 5 \\ +5 \\ \hline \end{array}$ **19.** $\begin{array}{r} 8 \\ +4 \\ \hline \end{array}$ **20.** $\begin{array}{r} 8 \\ +7 \\ \hline \end{array}$

21. $\begin{array}{r} 6 \\ +7 \\ \hline \end{array}$ **22.** $\begin{array}{r} 4 \\ +3 \\ \hline \end{array}$ **23.** $\begin{array}{r} 6 \\ +9 \\ \hline \end{array}$ **24.** $\begin{array}{r} 3 \\ +8 \\ \hline \end{array}$

Use Data from a Picture

Use the pictures to solve the problems.

1. How many bananas are there?

_____

2. How many pineapples are there?

_____

3. How many more oranges are there than watermelons?

_____

4. Are there more bananas or apples?

_____

5. How many pieces of fruit are there in all?

_____

6. Which fruit is the fewest in number? By how many?

_____ _____

7. Which type of fruit has the greatest number?

_____

8. How many fewer watermelons are there than oranges?

_____

9. Write the family of facts for Exercise 3.

_____

_____

10. Write the family of facts for Exercise 4.

_____

_____

Ordering Events

Order the following according to the rule.

1. Put in numerical order from least to greatest: 4, 9, 5, 2, 7.

_____

2. Put in numerical order from greatest to least: 6, 4, 3, 7, 1, 2.

_____

3. Put in order by size from smallest to largest: telephone, the world, mountain, house.

_____

_____

4. Put in order by loudness from quietest to loudest: jet engine, whisper, tractor, shout.

_____

_____

5. Put in alphabetical order: monkey, elephant, zebra, elk.

_____

_____

6. Put in order by age, oldest first: children, parents, great grandparents, grandparents.

_____

_____

Critical Thinking   Write these sentences in the correct order.

7. Anne drank a glass of water.
Anne was thirsty.
Anne got a glass from the cupboard.
Anne filled the glass.

_____

_____

_____

8. Scotty ran in the race.
Scotty won the race.
The race started!
Scotty warmed up near the track.

_____

_____

_____

Practice/**EXPLORING MATHEMATICS** © Scott, Foresman and Company/3

NAME

P9

SHARPEN YOUR SKILLS

Find a Pattern

About 900 years ago, I was made of animal bones tied to the bottom of shoes. What am I?

To find out, write the key sequence used in each exercise. Then connect the dots in the order of the answers given.

1. 4 8
    3 7
    5 9
    7 11

2. 5 2
    8 5
    3 0
    9 6

3. 5 12
    6 13
    8 15
    2 9

4. 5 3
    8 6
    6 4
    4 2

_____

_____

_____

_____

5. 11 6
    17 12
    6 1
    8 3

6. 10 2
    9 1
    12 4
    14 6

7. 9 15
    13 19
    18 24
    4 10

8. 5 8
    7 10
    16 19
    4 7

_____

_____

_____

_____

9. 14 5
    18 9
    9 0
    10 1

10. 7 1
    10 4
    9 3
    15 9

_____

_____

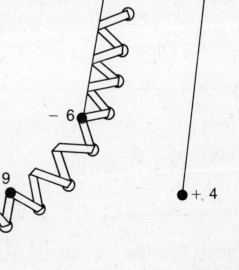

– 6

– 9

+ 3

+ 7

+ 6

– 8

+ 4

– 3

– 2

– 5

Understanding Hundreds

Use your place-value materials to make as many
tens as you can. Use **|** for 10 and **•** for 1 to record your work.

1. 70 ones

_____

2. 27 ones

_____

3. 45 ones

_____

4. 15 ones

_____

5. 93 ones

_____

_____

6. 64 ones

_____

_____

7. 63 ones

_____

_____

8. 85 ones

_____

_____

9. 30 ones

_____

10. 24 ones

_____

11. 53 ones

_____

12. 8 ones

_____

Use your place-value materials to make as many hundreds
as you can. Use **■** for 100, **|** for 10, and **•** for 1 to record your work.

13. 50 tens

_____

_____

14. 33 tens

_____

_____

15. 65 tens

_____

_____

16. 32 tens

_____

_____

17. 43 tens

_____

_____

18. 73 tens

_____

_____

19. 43 tens, 28 ones

_____

_____

20. 23 tens, 41 ones

_____

_____

21. 72 tens, 23 ones

_____

_____

Practice/EXPLORING MATHEMATICS © Scott, Foresman and Company/3

Hundreds, Tens, and Ones

Tell how many hundreds, tens, and ones.
Then write the number in standard form.

1.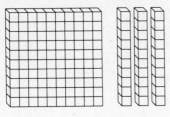

_____

2.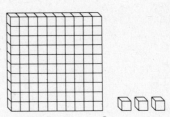

_____

3.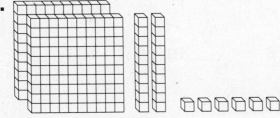

_____

4.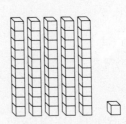

_____

Write each number in standard form.

5. 400 + 30 + 1 _____

6. 900 + 70 + 5 _____

7. 200 + 10 + 9 _____

8. 50 + 7 _____

9. three hundred eighty-four

10. seven hundred fifteen

_____

_____

In the following exercises, tell whether the 7 is the
hundreds digit, the tens digit, or the ones digit.

11. 607   **12.** 729   **13.** 437   **14.** 871

_____   _____   _____   _____

SHARPEN
YOUR
SKILLS

Comparing Numbers

Write < or > for each exercise.

1. 8 ◯ 4   **2.** 9 ◯ 7   **3.** 4 ◯ 13

4. 21 ◯ 9   **5.** 26 ◯ 64   **6.** 17 ◯ 20

7. 46 ◯ 219   **8.** 302 ◯ 79   **9.** 401 ◯ 608

10. 492 ◯ 261   **11.** 728 ◯ 816   **12.** 342 ◯ 380

13. 292 ◯ 216   **14.** 471 ◯ 476   **15.** 503 ◯ 504

Write < or > for each exercise.
Color the spaces with a < sign orange.
Color the spaces with a > sign brown.

19 ◯ 11

5 ◯ 9

3 ◯ 7

17 ◯ 8

29 ◯ 31

19 ◯ 12

80 ◯ 50

346 ◯ 338

67 ◯ 52

678 ◯ 674

201 ◯ 198

85 ◯ 77

10 ◯ 14

64 ◯ 66

999 ◯ 998

Practice/EXPLORING MATHEMATICS © Scott, Foresman and Company/3

Ordering Three-Digit Numbers

Write the numbers between

1. 144 and 151

2. 398 and 404

3. 987 and 993

_____   _____   _____

_____   _____   _____

4. 636 and 642

5. 209 and 203

6. 767 and 762

_____   _____   _____

_____   _____   _____

Write these numbers in order from least to greatest.

7. 109   55   86   68

8. 676   659   765   677

_____   _____

9. 789   879   987   798

10. 336   363   335   364

_____   _____

11. 204   402   240   420

12. 600   900   700   800

_____   _____

13. 542   245   452   425

14. 237   238   832   328

_____   _____

15. 735   537   375   753

16. 424   242   224   422

_____   _____

SHARPEN
YOUR
SKILLS

Patterns in Numbers

Count by twos.

1. Begin at 8. Count to 16.

2. Begin at 56. Count to 64.

_____

_____

Count by fives.

3. Begin at 15. Count to 35.

4. Begin at 30. Count to 50.

_____

_____

Describe each pattern. Then find the missing numbers.

5. 6, 11, 16, 21, _____, _____

6. 54, 64, 74, 84, _____, _____

_____

_____

7. 35, 37, 39, 41, _____, _____

8. 55, 57, 60, 62, 65, _____, _____

_____

_____

Count by tens to
connect the dots.
Start at the arrow.

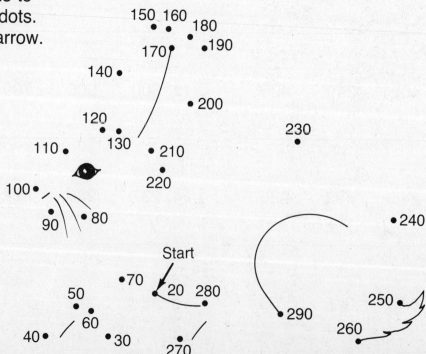

SHARPEN YOUR SKILLS

Rounding to Tens and Hundreds

What sea animals also go by the name of "sea-cows"?

Round each number to the nearest ten. Match each letter to its answer in the blanks below.

1. 654

_____ N

2. 896

_____ S

3. 122

_____ E

4. 980

_____ E

5. 582

_____ A

6. 607

_____ M

7. 345

_____ T

8. 989

_____ A

Round each number to the nearest hundred.

9. 399

_____ H

10. 454

_____ T

11. 790

_____ E

____ ____ ____
350   400   120

____ ____ ____ ____ ____ ____ ____ ____
610   580   650   990   500   980   800   900

Round each number to the nearest ten and then to the nearest hundred.

12. 603

_____

_____

13. 356

_____

_____

14. 298

_____

_____

15. 465

_____

_____

16. 476

_____

_____

17. 302

_____

_____

18. 873

_____

_____

19. 776

_____

_____

Use Logical Reasoning

The third grade is holding a track meet. Students are competing in many events. The following students are doing very well:

Students	Number
Jan	3
Trudy	6
James	5
Noah	7
Virginia	2
Fred	8

Use the clues to find each three-digit number by lining up the students' numbers as they appear when the winners are standing next to each other in first, second and third place. Name the students in order.

25-YARD DASH

1. Trudy's, Virginia's, and Fred's points make a number in which the tens and ones digits add up to the hundreds digit. The tens digit is greater than the ones digit.

862;
_____

LONG JUMP

2. Noah's, James's, and Jan's points make a number that is less than 600. The sum of the ones and hundreds digits is one more than the tens digit. The ones digit is greater than the hundreds digit and less than the tens digit.

_____

50-YARD DASH

3. Jan's, James's, and Fred's points make a three-digit number that is less than 400. The hundreds and the tens digits add up to the ones digit.

_____

QUARTER-MILE RUN

4. Trudy's, Jan's, and Noah's points make a three-digit number. The hundreds digit is 3 more than the tens digit. The ones digit is the largest number.

_____

Understanding Thousands

Write the word or words that will complete each sentence.

1. 30 ones = 3 _____

2. 50 tens = 5 _____

3. 80 hundreds = 8 _____

4. 60 tens = 6 _____

For each exercise, write the standard form.

5. 7,000 + 400 + 20 + 5

_____

6. six thousand, thirty-seven

_____

7. three thousand, six hundred
fifty-one

_____

8. Help Sir Bugalot to find the right
colors for his shield. Circle each
correct digit. Follow the chart to
color each shape.

Digit	Place Value	Color
1	thousands	Purple
3	hundreds	Blue
4	tens	Green
8	hundreds	Yellow
9	thousands	Brown
7	tens	Red

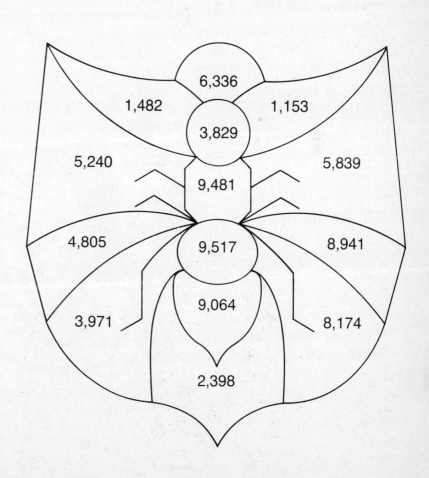

Comparing and Ordering Numbers

Write the numbers in order from least to greatest.

1. 5,438   5,727   4,648   _____

2. 6,320   6,294   5,930   _____

3. 4,051   4,016   4,091   _____

4. 3,208   3,824   3,842   _____

Write the numbers in order from greatest to least.

5. 2,306   3,504   2,241   _____

6. 4,516   3,961   4,821   _____

7. 7,319   7,621   7,341   _____

8. 5,206   5,381   5,260   _____

In each exercise, write < or >. Color each shape with a < yellow.
Color each shape with a > orange.

1,557 ◯ 1,602   3,904 ◯ 3,351   5,673 ◯ 5,637   6,026 ◯ 6,041

543 ◯ 430   6,839 ◯ 8,648   9,100 ◯ 8,999

Ten-Thousands and Hundred-Thousands

Write what the 7 in each number means.

1. 7,809 ____7 thousands____

2. 75,431 _____

3. 69,473 _____

4. 643,517 _____

5. 987,550 _____

6. 900,710 _____

7. 73,286 _____

8. 24,716 _____

9. 156,752 _____

10. 75,396 _____

11. 730,249 _____

12. 167,205 _____

Write each number in standard form.

13. ninety-eight thousand,
three hundred twenty-three

_____

14. six hundred thousand,
four hundred fifty

_____

15. forty-two thousand,
three hundred thirty-one

_____

16. twenty-two thousand,
four hundred ten

_____

17. five hundred three thousand,
six hundred nineteen

_____

18. nine hundred seventy thousand,
four hundred sixty

_____

19. three hundred fifteen thousand,
seven hundred sixty-eight

_____

20. forty-six thousand,
two hundred twelve

_____

SHARPEN
YOUR
SKILLS

Give Sensible Answers

Circle the most sensible answer.

1. The Hanson family enjoys music. How many radios do they have in the house?

   3   30   300

2. The circus is in town. The circus tent holds how many people?

   4   40   4,000

3. Roberto bought 3 game books. How much did he spend?

   $0.15 $15.00 $150.00

4. Clifton can lift twice his own weight. How many pounds can he lift?

   30   80   200

5. Julie is in the third grade. How many girls are in her scout troop?

   3   30   300

6. Lisa and Sara went to the movies. How many hours were they there?

   3   12   24

7. The zoo is open for the summer. The bear family had how many cubs?

   2   14   41

8. Iris went to the soccer game. How much did her ticket cost?

   $0.10 $2.00 $90.00

9. Tony plays on a basketball team. How many players are on the team?

   4   14   140

10. Ruby went to a concert. How many minutes long was the concert?

   2   20   100

11. Dan has a rowboat. How many people will it hold?

   5   50   500

12. Chan plays the cello. How much does he pay for an hour of lessons?

   $.10 $10 $100

Choose an Operation

Write the operation you need to use for each problem.
Then solve the problem.

1. There are 15 more stuffed bunnies on Hilda's table than on Theo's. Theo has 7. How many does Hilda have?

_____ _____

2. Paco is making braided key chains. If he makes 5 more, he will have 14. How many does he have?

_____ _____

3. Jennie is weaving belts. She made four yesterday. Today she will make 6 more than that. How many will she make today?

_____ _____

4. Kumiko makes paper fish. She made 3 in art class. How many will she need to make at home to have 11?

_____ _____

5. Nadia paints eggs and sells them. She wants to have 20 eggs to sell at a fair. She has 14. How many more will she need to have 20?

_____ _____

6. Philip makes sculptures by tying balloons together. He used 8 balloons yesterday. Today he used 11. How many has he used in all?

_____ _____

SHARPEN
YOUR
SKILLS

Renaming in Addition and Subtraction

Use your place-value materials. Make all the
new tens you can. How many tens and ones
will you have then?

1. Tens   Ones
   4   15

_____   _____

2. Tens   Ones
   1   17

_____   _____

3. Tens   Ones
   3   10

_____   _____

4. Tens   Ones
   3   12

_____   _____

5. Tens   Ones
   6   13

_____   _____

6. Tens   Ones
   2   11

_____   _____

Use your place-value materials. Trade 1 ten for 10 ones.
How many tens and ones will you have?

7. Tens   Ones
   9   1

_____   _____

8. Tens   Ones
   8   2

_____   _____

9. Tens   Ones
   5   4

_____   _____

Solve the problems.

10. Freida has 40 packs of seeds for sale.
If Susan gives her 12 more packs, how
many will she have? How many tens
and ones is this?

_____   _____

11. After Susan gave Freida the 12 packs,
Freida sold 6. How many did she
have left altogether? How many
tens and ones is this?

_____   _____

Seeds for Sale

Two-Digit Addition

Lizzy the bee was very busy. While gathering nectar, she became quite dizzy. Help dizzy Lizzy find her way home to the hive with no delay!

Work each exercise. Use your answers to give Lizzy a path through the maze. Follow the answers in the same order in which the exercises are numbered.

1. 29
   + 9

2. 36
   + 47

3. 52
   + 8

4. 48
   + 19

5. 16
   + 74

6. 26
   + 25

7. 39
   + 7

8. 17
   + 8

9. 47
   + 47

10. 29
    + 26

11. 54 + 18 = _____

12. 15 + 17 = _____

13. 66 + 7 = _____

14. 34 + 19 = _____

15. 45 + 36 = _____

16. 74 + 19 = _____

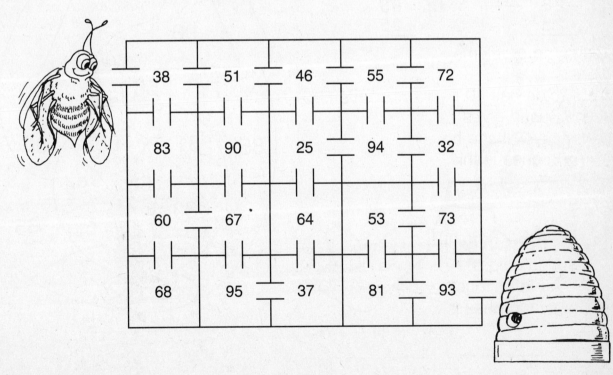

SHARPEN
YOUR
SKILLS

Addition: Mental Math

Find each sum mentally. Shade in the shape
for each answer.

1. 33
   +19

2. 28
   +26

3. 65
   +29

4. 78
   +17

5. 18
   +43

6. 36
   +46

7. 54
   + 8

8. 44
   +19

9. 57
   +26

10. 63
    +18

11. 28
    +28

12. 65
    +25

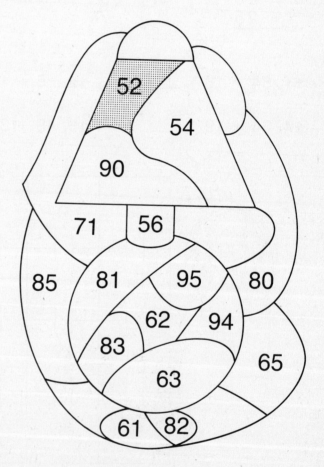

Three or More Addends

Find these sums.

1. 35 + 23 + 17

2. 61 + 72 + 8

3. 15 + 9 + 13

_____

_____

_____

4. 21 + 22 + 23

5. 14 + 90 + 2

6. 23 + 71 + 2

_____

_____

_____

Katie and her friends played a word game using
these letters. The chart shows the number of
points scored for each letter.

A	B	C	D	E	H	I	L	M	N	O	P	R	S	T	U
3	9	8	1	14	8	6	4	15	7	8	7	75	34	98	19

Add the values of the letters for each word below.
Circle the word with the highest score.

7. CAT

8. BET

9. EAT

10. BAR

_____

_____

_____

_____

11. ACT

12. ARE

13. ACE

14. RAT

_____

_____

_____

_____

15. DUEL

16. DART

17. BEAR

18. MAST

_____

_____

_____

_____

Two-Digit Subtraction

Find each difference. Work across and down. Check your
answer in the lower right corner by subtracting both
across and down.

1.

74	38	
46	29	

2.

61	47	
25	18	

3.

40	18	
23	9	

4.

82	46	
47	29	

5.

93	25	
56	17	

6.

53	27	
16	8	

Subtraction: Mental Math

Find each difference mentally. Then cross out the answer and the letter above it in the code below. Not all the answers will be crossed out. Write the remaining letters in order on the blanks at the bottom of the page to solve the riddle.

1. 29 − 12   **2.** 67 − 39   **3.** 92 − 24   **4.** 38 − 22

_____   _____   _____   _____

5. 64 − 33   **6.** 37 − 15   **7.** 98 − 56   **8.** 24 − 9

_____   _____   _____   _____

9. 42 − 16   **10.** 44 − 32   **11.** 88 − 39   **12.** 25 − 7

_____   _____   _____   _____

13. 56 − 37   **14.** 72 − 28   **15.** 65 − 38   **16.** 33 − 19

_____   _____   _____   _____

F	P	O	L	M	T	P	Y	E	R	U
29	12	68	33	26	44	14	89	31	17	22

P	K	A	F	I	P	T	E	L	Y	R
39	28	4	15	19	46	27	76	42	16	53

What is another name for air mail?

____ ____ ____   ____ ____ ____ ____

Estimation

Estimate each difference. Use front-end estimation.

| 1. $\begin{array}{r} 74 \\ -38 \\ \hline \end{array}$ | 2. $\begin{array}{r} 39 \\ -22 \\ \hline \end{array}$ | 3. $\begin{array}{r} 87 \\ -15 \\ \hline \end{array}$ | 4. $\begin{array}{r} 36 \\ -17 \\ \hline \end{array}$ |

Estimate each answer. Increase the first digit of each number by 1.

| 5. $\begin{array}{r} 53 \\ -18 \\ \hline \end{array}$ | 6. $\begin{array}{r} 88 \\ -45 \\ \hline \end{array}$ | 7. $\begin{array}{r} 22 \\ +39 \\ \hline \end{array}$ | 8. $\begin{array}{r} 27 \\ +64 \\ \hline \end{array}$ |

Tell whether the sum will be greater or less than 100.

9. $43 + 51$

10. $38 + 79$

11. $54 + 28$

12. $19 + 95$

_____

_____

_____

_____

13. $58 + 64$

14. $19 + 82$

15. $24 + 38$

16. $11 + 85$

_____

_____

_____

_____

Solve each problem by estimating the answer.

17. Carol and Sandy collected rocks to polish. Sandy collected 84 rocks. Carol collected 38 rocks. About how many more rocks did Sandy collect than Carol?

_____

18. Johnny had 98 baseball cards. He gave his friend, Bill, 36 cards to start his own collection. About how many baseball cards did Johnny have left?

_____

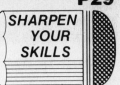

Deciding When an Estimate is Enough

Use the chart to answer the questions.

If you have 30¢, do you have enough money to buy the following items?

1. orange and apple

2. a bunch of carrots and a cucumber

3. cheese

_____

_____

_____

4. bread and milk

5. apple, banana, and carrots

6. cucumber and orange

_____

_____

_____

If you have 90¢, do you have enough money to buy the following items?

7. cheese, carrots, and an apple

8. lettuce and milk

9. cabbage, cheese and an orange

_____

_____

_____

Choose sets of at least 3 different items that you can buy for

10. 50¢

11. 87¢

12. 43¢

_____
or

_____
or

_____
or

_____

_____

_____

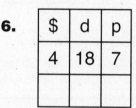

Using Money to Rename

Use your play money. Trade pennies for dimes and dimes
for dollars. Record your answers in the charts. Then write
out how many dollars, dimes, and pennies you have when you finish.

1.

$	d	p
2	14	12

_____

2.

$	d	p
0	12	5

_____

3.

$	d	p
1	15	10

_____

4.

$	d	p
5	16	18

_____

5.

$	d	p
2	9	25

_____

6.

$	d	p
4	18	7

_____

Make these sets of money with your play money. Trade
either a dime for 10 pennies or a dollar for 10 dimes. Record
how many dollars, dimes, and pennies you have when you finish.

7. Trade 1
dime for 10
pennies.

$	d	p
3	4	2

_____

8. Trade 1
dollar for
10 dimes.

$	d	p
1	2	5

_____

9. Trade 1
dime for 10
pennies.

$	d	p
4	7	12

_____

Exploring Three-Digit Addition

Use play money to show each amount of money in the place-value chart. Trade 10 pennies for 1 dime if you can. Then try to trade 10 dimes for 1 dollar. Write how much in all.

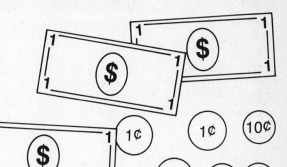

 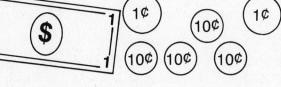

1.

$	d	p
1	4	3
+ 3	9	3

2.

$	d	p
2	5	6
+ 4	7	5

3.

$	d	p
0	7	9
+ 2	2	1

4.

$	d	p
3	2	4
+ 2	8	7

5.

$	d	p
3	6	4
+ 2	5	3

6.

$	d	p
5	2	1
+ 1	9	9

7.

$	d	p
4	7	4
+ 3	5	2

8.

$	d	p
5	0	2
+ 1	9	9

Solve each problem. You can use play money to help you.

9. Joanne went to the fair on Saturday and Sunday. On Saturday she spent $3.04 and on Sunday she spent $2.96. How much did she spend in all?

10. Bob and Ricardo were at the fair on Saturday. They each spent $2.88. How much did they spend in all?

Adding Three-Digit Numbers

Find each sum. **Remember** to estimate
before you find the exact answer.

1. 348
   +232

2. 596
   +101

3. 903
   +187

4. 237
   +355

5. 636
   +256

6. 345
   +219

7. 234
   +508

8. 423
   +249

9. 212
   +468

10. 607
    +318

11. 645 + 302 + 118 _____

12. 343 + 234 + 540 _____

13. 132 + 424 + 120 _____

14. 492 + 135 + 171 _____

15. 755 + 125 + 211 _____

16. 123 + 312 + 231 _____

Solve each problem.

17. Elvira, Joe, and Tammy drive
school buses. On Thursday
Elvira drove 121 students. Joe
drove 114 students. Tammy
drove 109 students. How many
students did they drive in all?

_____

18. On Friday they drove 124, 116,
and 110 students. How many
students did they drive in all on
Friday?

_____

Three- and Four-Digit Addition

Add.

1.
```
  4 6 9
+ 4 7 8
```

2.
```
  7 8 5
+ 1 3 7
```

3.
```
  8 4 8
+ 2 3 7
```

4.
```
  6 2 3
+ 7 4 8
```

5.
```
  3 8 5
+ 8 5 7
```

6.
```
  4 7 6
+ 8 5 6
```

7.
```
  9 4 8
+ 9 5 4
```

8.
```
  6 8 2
+ 6,5 7 3
```

9.
```
  2 5 6
  1 4 2
+ 3 5 8
```

10.
```
  7 2 7
  3 0 9
+ 4,2 4 5
```

11.
```
  4,3 8 6
  2,0 1 5
+ 7,2 4 8
```

12.
```
  3,2 9 8
  7,3 0 2
+ 7,2 3 6
```

13. 496 + 347 = _____

14. 4,601 + 299 = _____

15. 4,807 + 3,908 = _____

16. 7,856 + 1,691 = _____

Solve the problem.

17. For three days the campers collected pine cones. The first day they collected 375. The second day they collected 234. And on the third day they collected 296. How many pine cones were collected in three days?

_____

SHARPEN YOUR SKILLS

Choose a Computation Method

Use paper and pencil, mental math, or a calculator to find each answer. Write *P, M,* or *C* to show which method you used. Then draw a line from the problem to the correct answer.

1. $\begin{array}{r} 1\,0 \\ +\ \ 4 \\ \hline \end{array}$ 7 5

2. $\begin{array}{r} 3{,}4\,5\,4 \\ +3{,}7\,9\,8 \\ \hline \end{array}$ 2 4 3

3. $\begin{array}{r} 3\,8 \\ +2\,7 \\ \hline \end{array}$ 1,0 3 3

4. $\begin{array}{r} 1\,1\,8 \\ +1\,2\,5 \\ \hline \end{array}$ 7,2 5 2

5. $\begin{array}{r} 7\,6\,5 \\ +2\,6\,8 \\ \hline \end{array}$ 1 3 3

6. $\begin{array}{r} 4{,}2\,1\,1 \\ +3{,}9\,9\,9 \\ \hline \end{array}$ 6 1 0

7. $\begin{array}{r} 5\,0 \\ +2\,5 \\ \hline \end{array}$ 1 4

8. $\begin{array}{r} 3\,0\,0 \\ +4\,0\,0 \\ \hline \end{array}$ 8,2 1 0

9. $\begin{array}{r} 8\,7 \\ +4\,6 \\ \hline \end{array}$ 6 5

10. $\begin{array}{r} 5\,2\,7 \\ +\ \ 8\,3 \\ \hline \end{array}$ 7 0 0

Critical Thinking   Find the missing numbers.

11. $\begin{array}{r} 3\ 2\ \square \\ +\ 6\ \square\ 8 \\ \hline 9\ 8\ 3 \end{array}$

12. $\begin{array}{r} \square\ 3\ 7 \\ +\ 2\ 4\ \square \\ \hline 5\ 8\ 0 \end{array}$

13. $\begin{array}{r} 5\ 5\ 2 \\ +\ 6\ \square\ \square \\ \hline 1{,}2\ 4\ 0 \end{array}$

14. $\begin{array}{r} 3\ 9\ \square \\ +\ 2\ \square\ 9 \\ \hline 6\ 6\ 5 \end{array}$

Use Data from a Picture

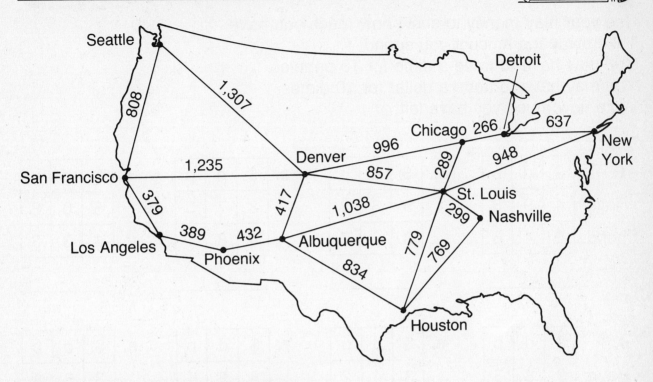

Use the map to solve each problem.

1. How many miles is it from San Francisco to St. Louis if you go through Denver?

_____

2. How many miles is it from Seattle to Los Angeles?

_____

3. How many miles is it from Phoenix to St. Louis? What city must you pass through?

_____

_____

4. What two smaller trips make up the total trip from Houston to Chicago? How many miles long is the trip?

_____

_____

5. How many miles is it from Denver to Nashville if you go through St. Louis?

_____

6. How many miles is it from Denver to New York by way of Chicago?

_____

Use after pages 124–125.

Exploring Three-Digit Subtraction

Use your play money to show how much you have.
Take away the amount you spend.
You may have to trade a dime for 10 pennies.
You may have to trade a dollar for 10 dimes.
Write how much you have left.

1.

$	d	p
4	9	5
2	4	6

Have:
Spend:

2.

$	d	p
5	8	2
3	6	7

3.

$	d	p
8	7	3
4	8	1

4.

$	d	p
5	0	8
1	3	3

5.

$	d	p
9	8	3
4	3	7

Have:
Spend:

6.

$	d	p
3	4	2
2	1	5

7.

$	d	p
8	9	4
3	9	6

8.

$	d	p
3	5	8
1	7	4

Number Sense   Make up a money subtraction problem
that makes you

9. trade a dollar for 10 dimes.

10. trade a dime for 10 pennies.

11. make no trades.

12. trade both a dollar for 10 dimes
and a dime for 10 pennies.

Subtract.

13. $7.82
      − $1.91

14. $2.03
       − $0.89

15. $6.54
       − $4.78

Subtracting Three-Digit Numbers

Find each difference. Remember to estimate before you find the exact answer. Then find your answer in the boxes below. Cross out the number and the letter in those boxes. The remaining letters will tell you the nickname for the state of Maine.

1. $\begin{array}{r}886\\-496\end{array}$	2. $\begin{array}{r}315\\-290\end{array}$	3. $\begin{array}{r}718\\-432\end{array}$	4. $\begin{array}{r}673\\-291\end{array}$
5. $\begin{array}{r}862\\-691\end{array}$	6. $\begin{array}{r}951\\-361\end{array}$	7. $\begin{array}{r}518\\-475\end{array}$	8. $\begin{array}{r}469\\-173\end{array}$
9. $\begin{array}{r}417\\-142\end{array}$	10. $\begin{array}{r}756\\-284\end{array}$	11. $\begin{array}{r}528\\-264\end{array}$	12. $\begin{array}{r}839\\-295\end{array}$
13. $\begin{array}{r}463\\-192\end{array}$	14. $\begin{array}{r}912\\-840\end{array}$	15. $\begin{array}{r}736\\-374\end{array}$	16. $\begin{array}{r}217\\-173\end{array}$

T	O	H	J	X	E	S		P	Q	R	I	N	E
19	171	904	275	43	19	390		575	72	286	67	309	486

T	Y	L	R	S	E	U	E
45	44	362	856	472	329	544	37

Q	S	I	T	R	A	O	T	M	E	N
590	312	382	287	25	55	296	123	264	444	271

Zeros in Subtraction

Find each difference.

1. 590
   −207

2. 204
   − 47

3. 400
   −328

4. 560
   −408

5. 103
   − 22

6. 100
   − 28

7. 430
   −120

8. 903
   − 85

9. 309
   −205

10. 240
    −107

11. 900
    −400

12. 308
    −107

13. 200
    − 44

14. 802
    −650

15. 204
    −129

Mixed Practice Use paper and pencil or
mental math to find each answer. Then write
which method you used.

16. 500 − 200

17. 398 + 243

18. 350 − 20

19. 457 + 345

_____   _____   _____   _____

_____   _____   _____   _____

20. 600 − 60

21. 873 − 261

18. 317 − 128

23. 380 + 620

_____   _____   _____   _____

_____   _____   _____   _____

Three- and Four-Digit Subtraction

Subtract. Write each letter above its matching answer below.
You will learn the state capitals of New York, Texas, and Virginia.

1. $\begin{array}{r} 4,007 \\ -2,136 \\ \hline \end{array}$ _____H

2. $\begin{array}{r} 387 \\ -\ 124 \\ \hline \end{array}$ _____T

3. $\begin{array}{r} 2,009 \\ -\ 322 \\ \hline \end{array}$ _____B

4. $\begin{array}{r} 560 \\ -\ 135 \\ \hline \end{array}$ _____Y

5. $\begin{array}{r} 5,006 \\ -2,065 \\ \hline \end{array}$ _____I

6. $\begin{array}{r} 274 \\ -\ 109 \\ \hline \end{array}$ _____A

7. $\begin{array}{r} 3,851 \\ -1,765 \\ \hline \end{array}$ _____N

8. $\begin{array}{r} 7,003 \\ -\ 392 \\ \hline \end{array}$ _____M

9. $\begin{array}{r} 3,201 \\ -1,122 \\ \hline \end{array}$ _____N

10. $\begin{array}{r} 534 \\ -\ 390 \\ \hline \end{array}$ _____A

11. $\begin{array}{r} 1,006 \\ -\ 251 \\ \hline \end{array}$ _____S

12. $\begin{array}{r} 7,965 \\ -4,343 \\ \hline \end{array}$ _____C

13. $\begin{array}{r} 2,354 \\ -1,030 \\ \hline \end{array}$ _____I

14. $\begin{array}{r} 769 \\ -\ 430 \\ \hline \end{array}$ _____N

15. $\begin{array}{r} 454 \\ -\ 129 \\ \hline \end{array}$ _____O

16. $\begin{array}{r} 409 \\ -\ 232 \\ \hline \end{array}$ _____L

17. $\begin{array}{r} 3,002 \\ -2,138 \\ \hline \end{array}$ _____R

18. $\begin{array}{r} 836 \\ -\ 109 \\ \hline \end{array}$ _____A

19. $\begin{array}{r} 3,212 \\ -2,133 \\ \hline \end{array}$ _____D

20. $\begin{array}{r} 3,030 \\ -\ 802 \\ \hline \end{array}$ _____U

The capital of New York is ____ ____ ____ ____ ____ ____.
727   177   1,687   165   339   425

The capital of Texas is ____ ____ ____ ____ ____ ____.
144   2,228   755   263   1,324   2,086

The capital of Virginia is ____ ____ ____ ____ ____ ____ ____ ____.
864   2,941   3,622   1,871   6,611   325   2,079   1,079

Use after pages 136–137.

Use Data from a Table

Use the data from one of the tables to solve each problem.

TRAIN SCHEDULE					
ARRIVALS			DEPARTURES		
From	Time	Name	To	Time	Name
BOSTON	8:18	CHIEF	CLEVELAND	6:30	CLOUD
DENVER	9:05	ROCKIES	LOS ANGELES	7:40	COAST
WASHINGTON, D.C.	9:56	CAPITAL	BALTIMORE	11:00	SUN
CHICAGO	10:55	EXPRESS	PHILADELPHIA	11:30	FLYER
NEW ORLEANS	11:56	SOUTHERN	SEATTLE	11:30	NORTHERN

1. At what time does the train arrive from New Orleans?

_____

2. At what time does the train leave for Baltimore?

_____

3. Where are the two trains going that leave at the same time?

_____

4. At what time does the train arrive from Washington, D.C.?

_____

5. What is the name of the train going to Seattle?

_____

6. What is the name of the train that arrives from Washington, D.C.?

_____

7. At what time does the train leave for Los Angeles? What is the name of the train?

_____

8. What is the name of the first train to leave? Where is it going?

_____

9. What is the name of the last train to arrive? From where does it arrive?

_____

10. What is the name of the train that arrives from Chicago? At what time does it arrive?

_____

Practice/EXPLORING MATHEMATICS © Scott, Foresman and Company/3

Telling Time: Nearest 5 Minutes and Minute

Write each time in two ways.

1.

_____

_____

2.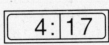

_____

_____

3.

_____

_____

Tell what time it will be. Write each time in two ways.

4.

10 minutes from now

_____

_____

5.

15 minutes from now

_____

_____

6.

30 minutes from now

_____

_____

Solve each problem. **Remember** that the short hand is the hour hand and the long hand is the minute hand.

7. On the clock, show a likely time for Susanne to eat dinner. Then write that time.

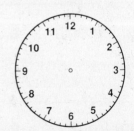

_____

8. Suppose it takes Susanne 30 minutes to eat dinner. Show when she will finish.

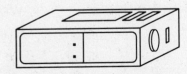

_____

Use after pages 150–153.

Reading Calendars: Days, Weeks, Months

Use the calendar to answer the questions.

MAY						
S	M	T	W	T	F	S
		1	2	3	4	5
6	7	8	9	10	11	12
13	14	15	16	17	18	19
20	21	22	23	24	25	26
27	28	29	30	31		

JUNE						
S	M	T	W	T	F	S
					1	2
3	4	5	6	7	8	9
10	11	12	13	14	15	16
17	18	19	20	21	22	23
24	25	26	27	28	29	30

JULY						
S	M	T	W	T	F	S
1	2	3	4	5	6	7
8	9	10	11	12	13	14
15	16	17	18	19	20	21
22	23	24	25	26	27	28
29	30	31				

1. What date is one week after June 12?

_____

2. What date is one week before May 20?

_____

3. What date is two weeks before July 24?

_____

4. On what day of the week does May 12 fall?

_____

Use the calendar to answer the questions.

FEBRUARY						
S	M	T	W	T	F	S
			1	2	3	4
5	6	7	8	9	10	11
12 Lincoln's Birthday	13	**14** Valentine's Day	15	16	17	18
19	**20** Presidents' Day	21	**22** Washington's Birthday	23	24	25
26	27	28				

5. What is the date of Washington's Birthday?

_____

6. On what day of the week is Presidents' Day?

_____

7. How many days are there from Lincoln's Birthday to Valentine's Day?

_____

8. On what dates does the first full week of February begin and end?

_____

Customary Lengths: Inches, Feet

Measure each object to the nearest half-inch.

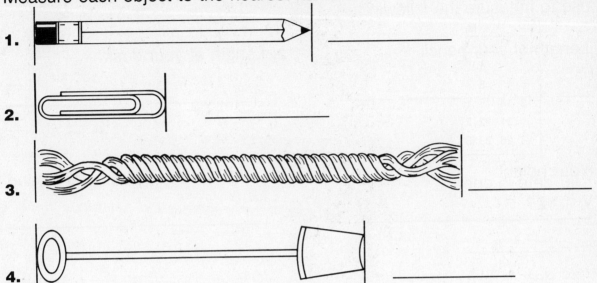

1. _____

2. _____

3. _____

4. _____

Measure the distance between each dot to the nearest half-inch. Then draw a line between the dots. Mark the measurement on the lines below.

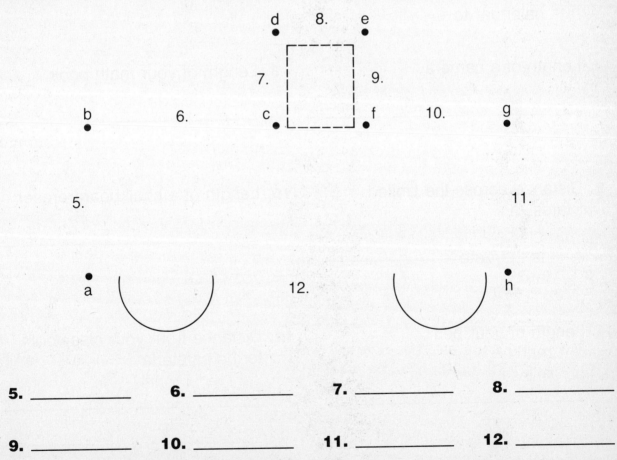

5. _____   6. _____   7. _____   8. _____

9. _____   10. _____   11. _____   12. _____

SHARPEN
YOUR
SKILLS

Customary Lengths: Yards, Miles

Write *inches, feet, yards,* or *miles* to tell which you
will use to measure the following.

1. Length of your pencil

_____

2. Length of your desk

_____

3. Your height

_____

4. Length of the school playground

_____

5. Distance from Hawaii to New York

_____

6. Length of an airplane

_____

7. Length of a camera

_____

8. Length of your math book

_____

9. Distance across the United States

_____

10. Length of a chalkboard eraser

_____

11. Length of your bed

_____

12. Distance from your classroom to the cafeteria

_____

Metric Lengths: Centimeters, Decimeters

Measure the ringmaster to the nearest centimeter.

1. About how long is his arm? _____

2. About how high is his boot? _____

3. About how tall is the ringmaster? _____

4. About how high is his hat? _____

5. About how wide is his hat? _____

6. About how long is his coat? _____

7. About how wide is his boot? _____

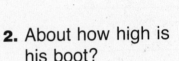

Problem Solving

8. How much higher is the ringmaster's hat than his boot?

_____

9. How much taller is the ringmaster than his coat?

_____

10. How much longer is the ringmaster's coat than his arm?

_____

11. How much wider is the ringmaster's hat than his boot?

_____

Metric Lengths: Meters, Kilometers

Think of four things you would measure using
each unit of measure below.

centimeters **1.** _____ **2.** _____

3. _____ **4.** _____

meters **5.** _____ **6.** _____

7. _____ **8.** _____

kilometers **9.** _____ **10.** _____

11. _____ **12.** _____

Mixed Practice Solve each problem.

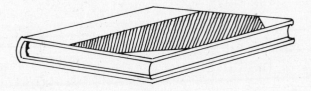

13. Sarah walks to school each day.
She walks 1,000 meters. How
many kilometers does she walk?

14. Frederico's book is 12 inches
long. How many feet long is his
book?

_____

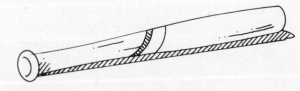

15. Tammy takes a giant step. It is
3 feet long. How many yards is
her giant step?

16. Bobby's baseball bat is 1 meter
long. How many centimeters
long is it?

_____

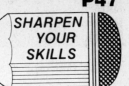

Make a Table

Use the four clues given to answer each question.

1. Thomas, Pauline, Carey, and Gale are eating lunch together.
Each child is eating a different food. The foods are pizza,
chicken, yogurt, and a taco. Which food is each child eating?

Clue 1: No child's name begins with the same letter as the
name of the food he or she is eating.
Clue 2: No child's name has the same number of letters as
the food he or she is eating.
Clue 3: Carey did not eat a taco, and Gale did not eat chicken
or yogurt.
Clue 4: Thomas did not eat pizza, and Pauline did not eat
yogurt.

	Pizza	Chicken	Taco	Yogurt
Thomas				
Gale				
Carey				
Pauline				

2. Julia, Anita, Ralph, and Keith were born in the same year.
They were born in May, June, July, and August.

Which one is the oldest? _____

Clue 1: Anita was born before Ralph.   Clue 2: Keith was born in July.
Clue 3: Ralph is the youngest.   Clue 4: Julia is older than Anita.

	May	June	July	August
Julia				
Anita				
Ralph				
Keith				

Perimeter

Use paper and pencil or mental math to find
the perimeter of each figure.

1.

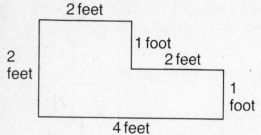

2.

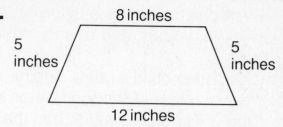

3.

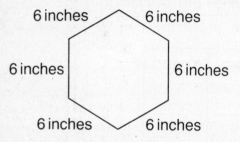

4.

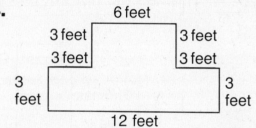

5.

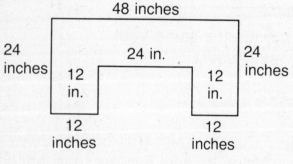

6.

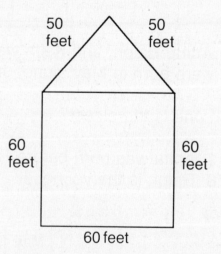

Counting Squares to Find Area

Find the area of each figure in square centimeters.

1.

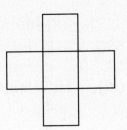

_____

2.

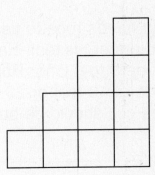

_____

3.

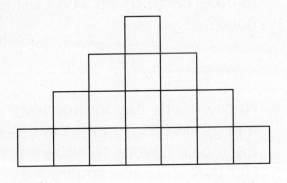

_____

4.

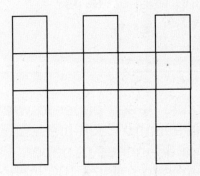

_____

Solve each problem.

5. Find the area in square centimeters of five pencils placed side by side. (Use grid paper.)

_____

6. Draw a figure with straight sides on a piece of grid paper. Then find its area.

_____

7. Which has the greater area, Figure A or Figure B? How much greater?

_____

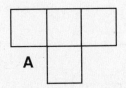

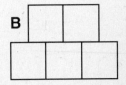

Deciding When an Estimate Is Enough

Write whether you need to find an exact answer or
an estimate. Then write the answer to each question.

1. Jason needs rope to tie down the
4 corners of his tent. Each piece
must be 3 feet long. Rope comes
in coils of 10 feet or 20 feet.
Which coil should he buy?

_____ _____

2. Elaine is buying fabric for
3 dresses. Each dress uses
4 yards of fabric. Fabric comes
cut to order. How much should
she buy?

_____ _____

3. There are 14 people on a
camping trip. They have 6 tents.
Each holds 2 people. Did they
bring enough tents?

_____ _____

4. Larry needs wood for a picture
frame. The frame is 24 inches on
each side. Framing wood is cut
to order. How much wood will he
need?

_____ _____

5. Ursula needs paper to wrap
6 presents. Each present will
take 33 inches of paper. Paper
comes in rolls of 100 inches and
250 inches. Which roll should
she buy?

_____ _____

6. Nancy needs hay for her lawn
and garden. Hay comes in bales.
Each bale covers 10 square feet.
Her garden is 100 square feet.
How many bales will she need?

Practice/EXPLORING MATHEMATICS © Scott, Foresman and Company/3

Temperature

Write the temperature in °C or °F.

1. 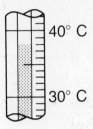 40° C

30° C

_____

2. 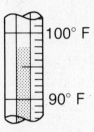 100° F

90° F

_____

3. 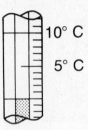 10° C

5° C

_____

4. 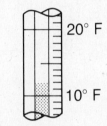 20° F

10° F

_____

Choose the more sensible answer. Put a check
on the line next to it.

5. Temperature for ice skating

30°F _____   70°F _____

6. Temperature for a picnic

28°C _____   75°C _____

7. Temperature of milk

5°C _____   40°C _____

8. Temperature of a healthy
person

98°F _____   78°F _____

Use a crayon to mark the correct temperature on each thermometer.

9. 29°F 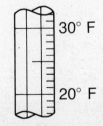 30° F

20° F

10. 72°C 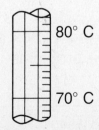 80° C

70° C

Use after pages 174–175.

Joining Groups of Equal Size

Tell how many.

1. $3 + 3 + 3 + 3 + 3 + 3 =$ _____

2. $2 + 2 + 2 + 2 + 2 + 2 =$ _____

3. $6 + 6 + 6 + 6 =$ _____

4. $8 + 8 + 8 + 8 + 8 =$ _____

5. $7 + 7 + 7 =$ _____

6. $5 + 5 + 5 + 5 =$ _____

7. $4 + 4 + 4 + 4 + 4 + 4 =$ _____

8. $9 + 9 + 9 + 9 =$ _____

9. $10 + 10 + 10 + 10 + 10 =$ _____

10. $15 + 15 + 15 + 15 + 15 =$ _____

11. $50 + 50 + 50 + 50 =$ _____

12. $43 + 43 + 43 =$ _____

13. $21 + 21 + 21 =$ _____

14. $13 + 13 + 13 =$ _____

15. $11 + 11 + 11 + 11 =$ _____

16. $12 + 12 + 12 =$ _____

17. $40 + 40 + 40 =$ _____

18. $100 + 100 =$ _____

Mixed Practice   Circle the one that is greater.

19. $6 + 3$ or $7 + 4$

20. $9 - 3$ or $9 - 4$

21. $6 - 3$ or $8 - 3$

22. $15 - 6$ or $14 - 6$

23. $35 + 35$ or $33 + 33$

24. $30 + 10$ or $40 - 10$

25. $21 - 9$ or $9 + 15$

26. $12 + 7$ or $8 + 10$

27. $28 + 28$ or $55 - 30$

28. $9 + 9$ or $36 - 12$

Critical Thinking   Look at Exercises 19–28.
Which of these problems can you do mentally?

_____

_____

Meaning of Multiplication: Repeated Addition

Write an addition and a multiplication sentence for each exercise.
Remember, use objects to help you.

1.

2.

3.

_____

_____

_____

_____

_____

_____

4.

5. (cats arranged in rows)

_____

_____

_____

_____

Estimation Use what you know about estimating sums
to estimate multiplication. Estimate each answer.

6. 19 + 19 + 19 = _____

7. 31 + 31 + 31 = _____

8. 12 × 3 = _____

9. 2 × 299 = _____

10. 3 × 107 = _____

11. 4 × 41 = _____

12. 309 × 2 = _____

13. 71 × 3 = _____

14. 79 × 4 = _____

SHARPEN
YOUR
SKILLS

Exploring Multiplication Using Arrays

Make an array and write a multiplication sentence
for each exercise. Use x's to make each array.

1. 3 twos

2. 3 fours

3. 2 + 2 + 2 + 2

_____

_____

_____

4. 4 rows of 5

5. 5 rows of 4

6. 7 sevens

_____

_____

_____

7. 8 rows of 2

8. 5 + 5 + 5

9. 6 sixes

_____

_____

_____

Problem Solving Make an array to
help solve the problem.

10. Red Rose Clocks displays 6 clocks
on each of 4 shelves. How many
clocks are displayed in all?

_____

Multiplication with Fives

Write two multiplication sentences for each picture.

1.

2.

3.

_____

_____

_____

_____

_____

_____

Multiply.

4.   4
    × 5

5.   2
    × 5

6.   8
    × 5

7.   5
    × 5

8.   7
    × 5

9.   9
    × 5

10.   3
    × 5

11.   6
    × 5

12.   5
    × 3

13.   5
    × 4

14.   5
    × 5

15.   5
    × 8

16. $5 \times 9 =$ _____

17. $5 \times 6 =$ _____

18. $5 \times 2 =$ _____

19. $5 \times 7 =$ _____

20. $5 \times 3 =$ _____

21. $5 \times 8 =$ _____

Use after pages 194–195.

SHARPEN YOUR SKILLS

Twos: Doubling

Find each answer.

1. xx   xx
   xx
xx   xx

$5 \times 2 =$ _____

2. xx xx xx xx
   xx xx

$6 \times 2 =$ _____

3. x x x x
x x x x

$4 \times 2 =$ _____

4. x x x x x x x x x
x x x x x x x x x

$2 \times 9 =$ _____

5. x x x x x x x x x
x x x x x x x x x

$2 \times 8 =$ _____

6. x x x x x x
x x x x x x

$2 \times 6 =$ _____

7. x x
x x

$2 \times 2 =$ _____

Find the missing number. Fill in each ring with the correct number of x's.

8. $2 \times \square = 14$

9. $2 \times \square = 8$

10. $2 \times \square = 18$

11. $2 \times \square = 6$

Draw a Picture

Read the paragraph. Draw pictures to help
answer the questions.

Marvin and I walk down a long hallway. One of the walls
is solid, but the other is divided into **2** panels. Each
panel contains **1** window. Each window is divided into
9 panes of glass. There is **1** plant on each window ledge.
Each plant has **2** leaves, and on each leaf are
2 caterpillars. Each caterpillar has **8** legs.

1. How many panes of glass are there all together? _____

2. How many leaves are there all together? _____

3. How many caterpillars are there all together? _____

4. How many caterpillar legs are there all together? _____

Use after pages 200–201.

Threes: Adding to a Known Fact

Why is this man like this animal?

To find out, multiply. Each time an answer is given
in the code, write the letter for that exercise.

1. $\begin{array}{r} 1 \\ \times\, 3 \end{array}$ B

2. $\begin{array}{r} 3 \\ \times\, 5 \end{array}$ A

3. $\begin{array}{r} 3 \\ \times\, 3 \end{array}$ U

4. $\begin{array}{r} 2 \\ \times\, 3 \end{array}$ E

5. $\begin{array}{r} 8 \\ \times\, 3 \end{array}$ F

6. $\begin{array}{r} 7 \\ \times\, 3 \end{array}$ K

7. $\begin{array}{r} 3 \\ \times\, 9 \end{array}$ J

8. $\begin{array}{r} 6 \\ \times\, 5 \end{array}$ I

9. $\begin{array}{r} 2 \\ \times\, 2 \end{array}$ H

10. $\begin{array}{r} 7 \\ \times\, 2 \end{array}$ G

11. $\begin{array}{r} 1 \\ \times\, 5 \end{array}$ P

12. $\begin{array}{r} 3 \\ \times\, 8 \end{array}$ O

13. $\begin{array}{r} 9 \\ \times\, 3 \end{array}$ N

14. $\begin{array}{r} 2 \\ \times\, 5 \end{array}$ M

15. $\begin{array}{r} 4 \\ \times\, 5 \end{array}$ L

16. $\begin{array}{r} 1 \\ \times\, 2 \end{array}$ T

17. $\begin{array}{r} 2 \\ \times\, 4 \end{array}$ S

18. $\begin{array}{r} 6 \\ \times\, 3 \end{array}$ R

19. $\begin{array}{r} 2 \\ \times\, 1 \end{array}$ Q

20. $\begin{array}{r} 4 \\ \times\, 3 \end{array}$ C

| 3 | 6 | 12 | 15 | 9 | 8 | 6 | | 4 | 6 | | 4 | 15 | 8 |

| 3 | 15 | 18 | 6 | | 20 | 6 | 14 | 8 |

Use after pages 202–203.

Fours: Doubling a Known Fact

$\begin{array}{r} 4 \\ \times\ 3 \\ \hline 12 \end{array}$ is the same as $\begin{array}{r} 2 \\ \times\ 3 \\ \hline 6 \end{array} + \begin{array}{r} 2 \\ \times\ 3 \\ \hline 6 \end{array} = 12$

Find each product using doubling.

1. $\begin{array}{r} 5 \\ \times\ 4 \\ \hline \end{array}$ is the same as

2. $\begin{array}{r} 4 \\ \times\ 4 \\ \hline \end{array}$ is the same as

3. $\begin{array}{r} 4 \\ \times\ 8 \\ \hline \end{array}$ is the same as

4. $\begin{array}{r} 2 \\ \times\ 4 \\ \hline \end{array}$ is the same as

5. $\begin{array}{r} 9 \\ \times\ 4 \\ \hline \end{array}$ is the same as

Find the missing factor. Fill in each ring with
the correct number of Xs.

6. $4 \times$ _____ $= 12$

7. $4 \times$ _____ $= 20$

8. $4 \times$ _____ $= 36$

Too Much Information

Use the information from this page of a catalog to complete the order form.

Men's Hiking Boots
Colors: tan, black, brown
Sizes: 6-13
Widths: A, B, C, D, E, EE
Price: $42
#M-44321

Women's Hiking Boots
Colors: tan, black, brown
Sizes: 6-11
Widths: A, B, C, D
Price: $38
#W-44321

Men's Running Shoes
Colors: black/red; black/green
Sizes: 6-12
Widths: A, B, C, D
Price: $39
#M-R3345

Women's Running Shoes
Colors: black/red; black/green
Sizes: 6-11
Widths: A, B, C, D
Price: $29
#W-R3345

Shoelaces
2 per package; Colors: black, brown, white
Price: $2
#55670

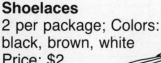

Shoe Polish
Colors: black, brown, white
Price: $3 per can
#77010

Name of Item	Catalog #	Color	Size	Quantity	Price Each	Total
	M-44321	brown	13E	2	$42	$84
Women's hiking boots		tan	7B	3		
	W-R3345	blk/red	6A	1		
	M-R3345	blk/green	9C	2		
	55670	white	X	6	2	
Shoe polish		brown	X	4	3	
					Total:	

NAME

Zero and One in Multiplication

Help the turtle get back to its shell. Multiply.
Then follow the path where your answer is 0 or 1.

2 × 3	6 × 2	8 × 4	5 × 5		
3 × 5	4 × 3	2 × 9	8 × 2		
3 × 4	8 × 3	7 × 3	0 × 5	1 × 1	8 × 5
9 × 2	3 × 5	4 × 5	0 × 1	5 × 3	2 × 4
4 × 2	6 × 3	8 × 5	0 × 5	7 × 5	4 × 3
8 × 3	5 × 5	3 × 0	0 × 2	6 × 5	9 × 2
9 × 4	4 × 3	8 × 0	7 × 2	7 × 4	2 × 3
3 × 5	3 × 4	0 × 1	5 × 0	0 × 5	9 × 0
		5 × 3	8 × 4	9 × 3	8 × 0
		0 × 3	6 × 0	1 × 1	0 × 5

Using Order to Multiply with 6 Through 9

Write two multiplication sentences for each picture.

1.

2.

3.

_____

_____

_____

_____

_____

_____

Write each product. Use the order property to help you.

4.	**5.**	**6.**	**7.**	**8.**
6	7	4	3	9
×4	×3	×8	×5	×5

Mental Math Find each answer.

9.	**10.**	**11.**	**12.**	**13.**
5	4	3	2	1
×5	×4	×8	×2	×9

Solve the problem.

14. During a baseball game, the
visiting team scored 6 runs in
each of the first 4 innings. How
many runs had they scored
after 4 innings?

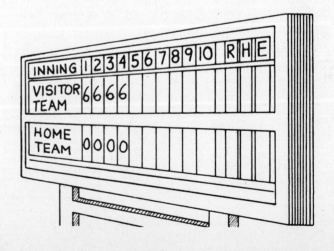

_____

Multiplication with 6 Through 9 on the Number Line

Find each product. Remember to use a number line if you need it. In the box at the right, find your answer. Cross out the letter across from it. The remaining letters will spell something you need. Write the word in the space at the bottom of the page.

Find each answer.

1.	9 $\times 3$	**2.**	7 $\times 4$	**3.**	8 $\times 3$	**4.**	2 $\times 8$
5.	6 $\times 3$	**6.**	4 $\times 9$	**7.**	5 $\times 8$	**8.**	5 $\times 6$
9.	1 $\times 5$	**10.**	2 $\times 7$	**11.**	5 $\times 9$	**12.**	3 $\times 7$
13.	7 $\times 1$	**14.**	3 $\times 5$	**15.**	2 $\times 4$	**16.**	4 $\times 3$
17.	5 $\times 5$	**18.**	5 $\times 2$	**19.**	3 $\times 2$	**20.**	4 $\times 8$
21.	5 $\times 4$	**22.**	7 $\times 5$	**23.**	0 $\times 6$	**24.**	3 $\times 3$

What do all boys and girls need plenty of?

_____

A	35
L	7
L	0
B	32
O	18
Y	6
S	79
A	21
N	12
D	20
G	27
I	14
R	36
L	19
S	30
N	24
E	47
E	51
D	28
P	26
L	45
E	15
N	5
T	9
Y	8
O	40
F	16
I	25
T.	10

Choose an Operation

Use the chart below to solve each problem. Write whether you add, subtract, or multiply. Then find each answer.

Amount Spent on Clothes and Supplies for School

	Friday	Saturday	Sunday	Monday	Total
Joey	$3	$6	$2	$7	$18
Linda	$4	$5	$8	$8	$25
Maurice	$8	$3	$10	$3	
Jasmine	$2	$4	$5	$7	$18
Total		$18	$25		$85

1. How much did Maurice spend in all?

_____

2. How much did Linda spend on Sunday and Monday?

_____

3. How much more did Maurice spend than Jasmine on Sunday?

_____

4. What is the total that Maurice spent on Saturday and Monday?

_____

5. How much more did Joey spend on Saturday than on Friday?

_____

6. What was the total amount spent on Friday?

_____

7. How much more did Linda spend than Joey?

_____

8. How much did Joey and Jasmine spend on Monday?

_____

9. On two of the days the same amount was spent. How much was it?

_____

10. How much more was spent on Sunday than on Saturday?

_____

Square Products

Find each product. Match each letter, except *s*,
to its answer in the blanks below.

What is raised in wet countries?

1. $3 \times 3 =$ **2.** $5 \times 5 =$ **3.** $9 \times 9 =$ **4.** $2 \times 2 =$

     R   M   L   E

5. $6 \times 6 =$ **6.** $8 \times 8 =$ **7.** $1 \times 1 =$ **8.** $4 \times 4 =$

     L   A   B   U

                                                          S

$\overline{} \ \overline{} \ \overline{} \ \overline{} \ \overline{} \ \overline{} \ \overline{} \ \overline{}$
16   25   1   9   4   81   36   64

Critical Thinking   Make or draw arrays to help you solve each problem.

9. Write the next two numbers in this pattern.

3, 6, 9, 12, 15, _____ , _____

10. Write the next three numbers in this pattern.

4, 8, 12, _____ , _____ , _____

Using Tens to Multiply with 9

Fill in the blanks below.

1. $10 \times 9 = 90$

$9 \times 9 =$ _____ less than 90

$9 \times 9 =$ _____

2. $10 \times 8 = 80$

$9 \times 8 =$ _____ less than 80

$9 \times 8 =$ _____

3. $10 \times 7 = 70$

$9 \times 7 =$ _____ less than 70

$9 \times 7 =$ _____

4. $10 \times 6 = 60$

$9 \times 6 =$ _____ less than 60

$9 \times 6 =$ _____

5. $10 \times 5 = 50$

$9 \times 5 =$ _____ less than 50

$9 \times 5 =$ _____

6. $10 \times 4 = 40$

$9 \times 4 =$ _____ less than 40

$9 \times 4 =$ _____

7. $10 \times 3 = 30$

$9 \times 3 =$ _____ less than 30

$9 \times 3 =$ _____

8. $10 \times 2 = 20$

$9 \times 2 =$ _____ less than 20

$9 \times 2 =$ _____

9. $10 \times 1 = 10$

$9 \times 1 =$ _____ less than 10

$9 \times 1 =$ _____

Solve each problem.

10. Josie is planting a garden. She has 9 rows of tomatoes with 9 tomatoes in each row. How many does she have in all?

_____

11. Daniel has planted cucumbers. He has 5 rows. Each row has 9 cucumbers. How many does he have?

_____

The Last Three Facts: 6 × 7, 6 × 8, 7 × 8

Find each answer. Match the letter for each answer in the code below to find the answer to the following question:

What is the nickname for Texas?

1. 8 × 7 = _____
E

2. 9 × 3 = _____
T

3. 6 × 9 = _____
S

4. 5 × 7 = _____
A

5. 8 × 0 = _____
L

6. 4 × 8 = _____
T

7. 9 × 5 = _____
O

8. 8 × 8 = _____
H

Mixed Practice Use a calculator, paper and pencil, or mental math to find each answer.

9.   494
       − 55
         T

10.   6
         × 7
         N

11.   831
       + 207
         R

12.   9
         × 8
         T

13.   982
       − 409
         S

14.   237
       + 564
         A

15.   4
         × 9
         E

16. 5,692
       + 245
         E

_27_ _64_ _56_

_0_ _45_ _42_ _5,937_   _54_ _439_ _801_ _1,038_

_573_ _27_ _35_ _32_ _36_

SHARPEN YOUR SKILLS

Find a Pattern

Give the missing numbers.

1. Pattern: Add 2

$\underline{\quad 3 \quad}$ ___ ___ ___ ___ ___

2. Pattern: Add 4

$\underline{\quad 1 \quad}$ ___ ___ ___ ___ ___

3. Pattern: Add 2, add 3, add 4, add 5, add 6

$\underline{\quad 1 \quad}$ ___ ___ ___ ___ ___

4. Pattern: Add 2, add 4, add 8, add 16, add 32

$\underline{\quad 2 \quad}$ ___ ___ ___ ___ ___

5. Pattern: Add 2, subtract 1

$\underline{\quad 0 \quad}$ ___ ___ ___ ___ ___

6. Pattern: Subtract 5

$\underline{\quad 75 \quad}$ ___ ___ ___ ___ ___

7. Pattern: Add 5, subtract 4, add 6

$\underline{\quad 4 \quad}$ ___ ___ ___ ___ ___

Give the pattern.

8. 0 3 6 9 12 15

Pattern: _____

_____

9. 12 19 26 33 40

Pattern: _____

_____

10. 0 10 5 15 10 20

Pattern: _____

_____

11. 99 90 81 72 63

Pattern: _____

_____

12. 12 33 45 66 78 99

Pattern: _____

_____

13. 14 12 15 13 16 14

Pattern: _____

_____

14. 15 13 10 20 18 15 25

Pattern: _____

_____

Using Multiplication

Which buildings in Toytown are the same size?
Count the layers, rows, and blocks in each row.
Then find the product.

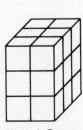
Feed Store

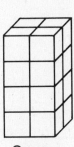
Grocery

School

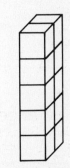
Meeting Hall

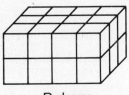
Bakery

Library

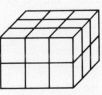
Courthouse

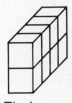
Firehouse

1. Feed Store

_____ × _____ × _____ = _____

2. Bakery

_____ × _____ × _____ = _____

3. Grocery

_____ × _____ × _____ = _____

4. Library

_____ × _____ × _____ = _____

5. School

_____ × _____ × _____ = _____

6. Courthouse

_____ × _____ × _____ = _____

7. Meeting Hall

_____ × _____ × _____ = _____

8. Firehouse

_____ × _____ × _____ = _____

The _____ and the _____ have the same number of blocks.

The _____ and the _____ have the same number of blocks.

Solid Figures

Write the name of the shape of each item. Use *cylinder,*
sphere, cube, or *rectangular prism.*

1. a beach ball

_____

2. a telescope

_____

3. a roll of paper towels

_____

4. an orange

_____

5. a shoebox

_____

6. a block

_____

Answer each question.

7. How many surfaces does a
sphere have?

_____

8. How many corners does a
cube have?

_____

Make a drawing of each of these shapes.

9. a rectangular prism

10. a cylinder

11. a cube

12. a sphere

Geometric Figures

The On-Time Train is made up of
geometric shapes.

1. Color the circles orange.

2. Color the triangles yellow.

3. Color the pentagons green.

4. Color the hexagons purple.

5. Color the octagons pink.

6. Outline the squares with red.

Shape	Number of each in the train	Number of corners of each
Circle		
Triangle		
Pentagon		
Hexagon		
Octagon		
Square		

7. Complete the table. Use the table to answer Exercise 8.

8. Which shape is used the greatest number of times?

_____

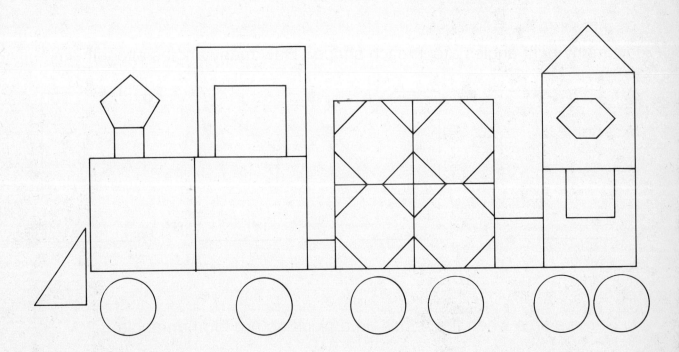

SHARPEN YOUR SKILLS

Angles

Is the angle a right angle? Mark *Yes* or *No*.

1.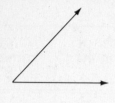

Yes   No

2.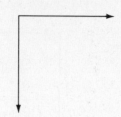

Yes   No

3.

Yes   No

4.

Yes   No

5.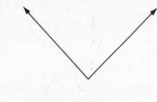

Yes   No

6.

Yes   No

How many right angles are in each shape? How many angles altogether?

7.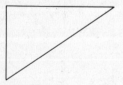

8.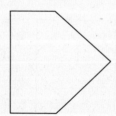

9.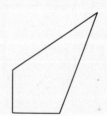

10.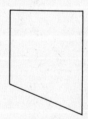

_____   _____   _____   _____

11. Name a time when the hands on a clock do not form an angle.

_____

Symmetry

In each row, mark the items that look as though they have a line of symmetry.

1.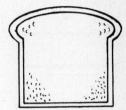

2.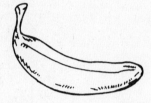

Does each figure look symmetric?
Mark *Yes* or *No.*

3.

Yes   No

4.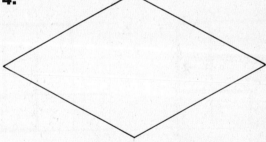

Yes   No

5.

Yes   No

6.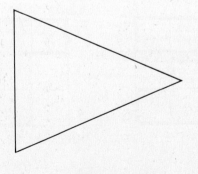

Yes   No

Use after pages 264–265.

Congruence

For each exercise tell whether each figure is
congruent with the first figure in the row. Mark *Yes*
or *No*.

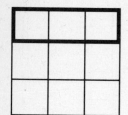

1.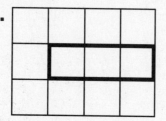

Yes   No

2.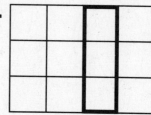

Yes   No

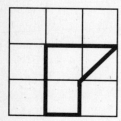

3.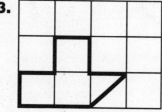

Yes   No

4.

Yes   No

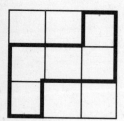

5.

Yes   No

6.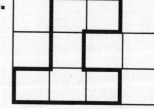

Yes   No

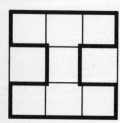

7.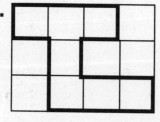

Yes   No

8.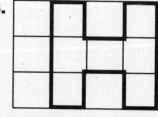

Yes   No

NAME

SHARPEN YOUR SKILLS

Use Data from a Graph

Student	Number of Times Swimming
Joe	⊗ ⊗ ⊗ ⊗
Freida	⊗ ⊗ ⊗ ⊗ ⊗
Lester	⊗
Marilyn	⊗ ⊗
Kris	⊗ ⊗ ⊗ ⊗ ⊗ ⊗

Each ⊗ equals one time.
The graph above shows how many times some students went swimming during a two-week period in July.

1. Who went swimming the least number of times?

____

2. Who went swimming the greatest number of times?

____

3. How many more times did Marilyn go swimming than Lester?

____

4. Which students went swimming more times than Joe?

____

5. What was the total number of times Joe and Freida went swimming?

____

6. How many more times did Kris go swimming than Joe?

____

NAME

Counting Cubes to Find Volume

SHARPEN YOUR SKILLS

Find the volume of each solid shape in cubic units.

1.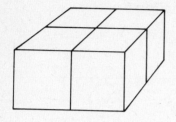

_____

2.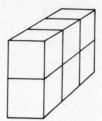

_____

3.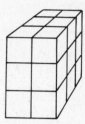

_____

4.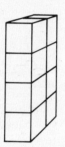

_____

5.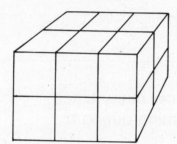

6.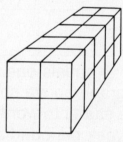

_____

7.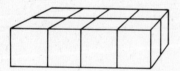

_____

8.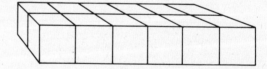

_____

9.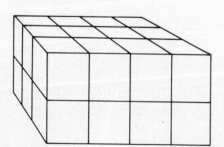

_____

10.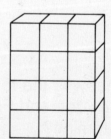

_____

Customary Units of Capacity

Mark the more sensible customary measure for each figure.

1.

Cup   or   quart

2.

Cup   or   gallon

3.

Pint   or   gallon

4.

Cup   or   quart

5.

Pint   or   gallon

6.

Cup   or   gallon

Mark the more sensible customary measure for each item.

7.	Bottle of shampoo	2 cups	2 gal
8.	Can of soup	1 cup	1 qt
9.	Washing machine	30 pt	30 gal
10.	Fish tank	5 pt	5 gal
11.	Glass of milk	1 pt	1 cup

Metric Units of Capacity

Write *liter* or *milliliter* to tell which you would
use to tell how much each item holds.

1.

_____

2.

_____

3.

_____

4.

_____

5. An aquarium _____

6. A fish pond _____

7. A tea cup _____

8. A teaspoon _____

9. A drinking fountain _____

Solve the problem.

10. Sandy's kitten drinks 75 mL of
milk each day. How much milk
will the kitten drink in 10 days?

_____

NAME

Customary Units of Weight

Write *ounces* or *pounds* to tell what you
would use to measure the weight of each item.

1.

2.

_____

_____

3. A full suitcase _____

4. A paintbrush _____

5. A banana _____

6. A sofa _____

7. A model plane _____

8. A car _____

9. An apple _____

10. An elephant _____

11. A baseball cap _____

12. A hair bow _____

13. A toothbrush _____

14. A large dog _____

15. A truck _____

16. A giraffe _____

17. A bowling ball _____

18. A comb _____

Use after pages 278–279.

SHARPEN YOUR SKILLS

Metric Units of Mass

Write *grams* or *kilograms* to tell what you
would use to measure the weight of each item.

1.

_____

2.

_____

3. A bowling ball _____   **4.** A toothbrush _____

5. A truck _____   **6.** A comb _____

7. A baseball cap _____   **8.** A football _____

9. Two marbles _____   **10.** A car _____

11. A kitchen chair _____   **12.** A ruler _____

13. A plastic dish _____   **14.** An elephant _____

15. A television _____   **16.** A large dog _____

17. A box of cereal _____   **18.** An apple _____

Too Little Information

Tell how you would solve each problem. If there is
missing information, tell what you need to know.

1. Each boat on the ride at Jay's
Amusement Park holds four
children. How many children
can take the ride at one time?

_____

_____

_____

_____

_____

2. The snack bar serves juice in
8-ounce and 12-ounce cups.
How many ounces of juice did
Ken and Kimo buy?

_____

_____

_____

_____

_____

3. Peter was baking a cake and
needed 8 ounces of raisins.
How many boxes of raisins
should he buy?

_____

_____

_____

_____

4. Jane and Aaron bought two
hamburgers each. How much
did they spend altogether?

_____

_____

_____

_____

Estimating Weights and Volumes

Is the correct volume shown for each picture?
Write *Yes* or *No*.

1.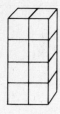

more than 8 cubic units

_____

2.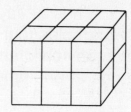

less than 14 cubic units

_____

3.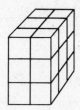

more than 20 cubic units

_____

4.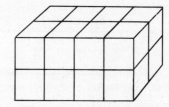

exactly 16 cubic units

_____

Tell how many of each object would make about 1 pound.

5. 8-ounce apple

_____

6. 3-ounce bag of nuts

_____

7. 4-ounce hamburger

_____

8. 2-ounce pen

_____

9. 16-ounce book

_____

10. 5-ounce banana

_____

Sharpen Your Skills

Sharing

Find how many in each group.

1. 16 in all
2 equal groups

2. 10 in all
5 equal groups

3. 12 in all
3 equal groups

_____

_____

_____

4. 20 in all
5 equal groups

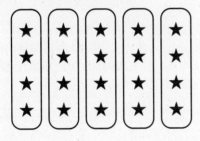

5. 18 in all
6 equal groups

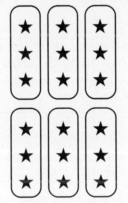

6. 28 in all
4 equal groups

_____

_____

_____

Solve each problem. Use the picture at the right.

7. How many apples are there? _____

8. How many groups of 7 can be made? _____

9. How many groups of 2 can be made? _____

10. If 7 equal groups are made, how
many will be in each group? _____

SHARPEN
YOUR
SKILLS

Grouping

Use small objects or draw pictures to find each answer.

1. 9 in all
3 in each group
How many groups?

_____

2. 24 in all
8 in each group
How many groups?

_____

3. 18 in all
2 in each group
How many groups?

_____

4. 30 in all
10 in each group
How many groups?

_____

5. 25 in all
5 in each group
How many groups?

_____

6. 10 in all
5 in each group
How many groups?

_____

7. 14 in all
7 in each group
How many groups?

_____

8. 15 in all
5 in each group
How many groups?

_____

9. 20 in all
4 in each group
How many groups?

_____

10. 24 in all
6 in each group
How many groups?

_____

11. 16 in all
8 in each group
How many groups?

_____

12. 36 in all
6 in each group
How many groups?

_____

Solve each problem.

13. Noreen has 15 pencils. She
wants to put them in pencil
cases. Each case holds 5
pencils. How many cases will
she need?

_____

14. Daria has 18 pencils. Her pencil
cases hold 6 pencils each. How
many cases does she have?

_____

Meaning of Division

Write the division sentence for each exercise.
Draw pictures to help you.

1. 20 in all
4 groups

[ ] in each group

$20 \div 4 =$ _____

2. 6 in all
2 in each group

[ ] groups

$6 \div 2 =$ _____

3. 8 in all
4 in each group

[ ] groups

$8 \div 4 =$ _____

4.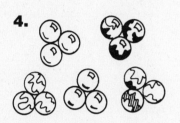

[ ] in all

5 groups

[ ] $\div 5 =$ _____

5.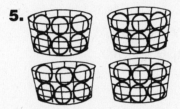

[ ] in all

4 baskets

[ ] $\div 4 =$ _____

6.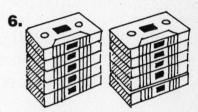

[ ] in all

2 stacks

[ ] $\div 2 =$ _____

7.

12 in all
4 blocks

[ ] piles

$12 \div 4 =$ _____

8.

12 in all
3 in each group

[ ] groups

$12 \div 3 =$ _____

9.

18 in all
3 in each group

[ ] groups

$18 \div 3 =$ _____

Use after pages 304–305.

Dividing by Two and Three

Write each answer.
Remember to draw pictures to help you.

1. $12 \div 3 =$ _____   **2.** $15 \div 3 =$ _____   **3.** $27 \div 3 =$ _____

4. $6 \div 2 =$ _____   **5.** $9 \div 3 =$ _____   **6.** $24 \div 3 =$ _____

7. $6 \div 3 =$ _____   **8.** $18 \div 2 =$ _____   **9.** $21 \div 3 =$ _____

10. $16 \div 2 =$ _____   **11.** $10 \div 2 =$ _____   **12.** $18 \div 3 =$ _____

To find two things you use every day,
divide. Then shade each shape where
the answer is 2, 4, 6, or 8.

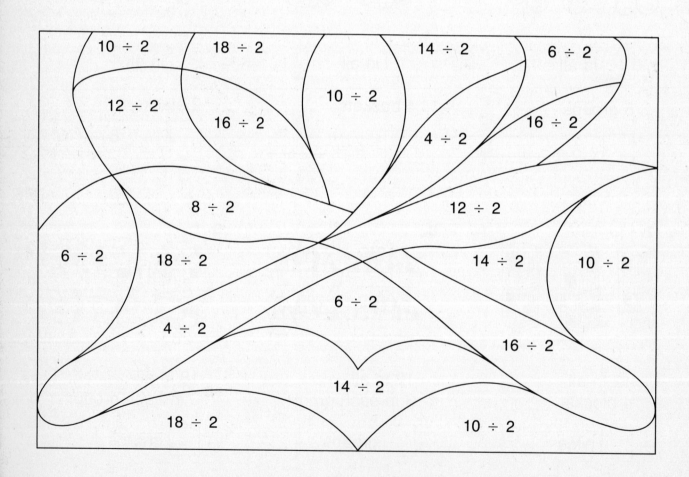

Dividing by Four

What has one horn and gives milk?

Find each quotient. Write each letter above its answer.

1. $12 \div 3 =$ _____ **L**   **2.** $18 \div 3 =$ _____ **K**   **3.** $20 \div 4 =$ _____ **T**

4. $36 \div 4 =$ _____ **U**   **5.** $16 \div 2 =$ _____ **C**   **6.** $9 \div 3 =$ _____ **I**

7. $8 \div 4 =$ _____ **M**   **8.** $4 \div 4 =$ _____ **K**   **9.** $14 \div 2 =$ _____ **R**

_____	_____	_____	_____	
2	3	4	1	

_____	_____	_____	_____	_____
5	7	9	8	6

Write each answer.

10. $16 \div 4 =$ _____   **11.** $36 \div 4 =$ _____   **12.** $24 \div 4 =$ _____

13. $24 \div 3 =$ _____   **14.** $40 \div 4 =$ _____   **15.** $28 \div 4 =$ _____

16. $12 \div 4 =$ _____   **17.** $18 \div 2 =$ _____   **18.** $32 \div 4 =$ _____

19. $4\overline{)20}$ _____   **20.** $4\overline{)8}$ _____   **21.** $4\overline{)36}$ _____

Try and Check

Solve each problem. Use try and check to find the answer.

1. Larry and Laurie have 75 baseball cards. Laurie has 15 more than Larry. How many cards does each of them have?

_____

2. There are 90 cars and trucks on a dealer's parking lot. There are 20 more cars than trucks. How many of each kind are there?

_____

3. Francis picked 36 apples. Some are bruised. There are 10 fewer bruised apples than good ones. How many of each kind are there?

_____

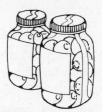

4. There are 24 jars of beans or olives on a shelf. There are 5 times as many jars of beans as of olives. How many jars of each are there?

_____

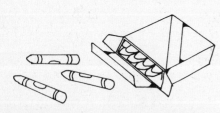

5. Joanna and Norman have 100 crayons. Norman has 14 more than Joanna. How many crayons does each of them have?

_____

6. Linda and Stacey have 21 magazines between them. Stacey has 2 times as many as Linda. How many magazines does each of them have?

_____

Dividing by Five

Write each quotient.

1. How many groups of 5?

2. How many in 5 equal groups?

$40 \div 5 =$ _____

$15 \div 5 =$ _____

3. $20 \div 5 =$ _____

4. $45 \div 5 =$ _____

5. $16 \div 4 =$ _____

6. $10 \div 2 =$ _____

7. $35 \div 5 =$ _____

8. $27 \div 3 =$ _____

9. $15 \div 5 =$ _____

10. $20 \div 4 =$ _____

11. $25 \div 5 =$ _____

12. $28 \div 4 =$ _____

13. $30 \div 5 =$ _____

14. $15 \div 3 =$ _____

15. $4\overline{)24}$

16. $5\overline{)40}$

17. $4\overline{)28}$

18. $5\overline{)45}$

19. $3\overline{)21}$

20. $4\overline{)32}$

21. $2\overline{)16}$

22. $3\overline{)24}$

23. $4\overline{)16}$

24. $3\overline{)3}$

25. $5\overline{)15}$

26. $3\overline{)18}$

27. $4\overline{)36}$

28. $3\overline{)12}$

29. $2\overline{)14}$

30. $5\overline{)5}$

Write a Number Sentence

Write a number sentence for each problem.
Then find the answer.

1. Four students each used 8 books
for information to write a fact
book about animals. How many
books were used in all?

_____

_____

2. There were only 9 pages in the
fact book. Four animals were
discussed on each page. How
many animals were in the fact
book?

_____

3. An elephant lives for about
60 years. A cat lives for about
13 years. How many more
years does an elephant live than
a cat?

_____

_____

4. The dromedary, or one-humped
camel, can easily travel
180 kilometers one day and
162 kilometers the next day. How
many kilometers can it travel in
two days?

_____

_____

5. An octopus has 8 arms. Each
arm has 9 suction cups to catch
and to hold food. How many
suction cups are there in all?

_____

_____

Families of Facts

Complete the number sentences. Write the number sentences to complete each family of facts.

1. $2 \times 3 =$ _____

$6 \div 2 =$ _____

_____

_____

2. $5 \times 6 =$ _____

$30 \div 5 =$ _____

_____

_____

3. $4 \times 7 =$ _____

$28 \div 4 =$ _____

_____

_____

4. $3 \times 3 =$ _____

_____

5. $4 \times 6 =$ _____

$24 \div 4 =$ _____

_____

_____

6. $3 \times 6 =$ _____

$18 \div 3 =$ _____

_____

_____

Write each answer. Circle the fact that does not belong to the family.

7. $32 \div 4 =$ _____

$7 \times 4 =$ _____

$28 \div 4 =$ _____

$4 \times 7 =$ _____

$28 \div 7 =$ _____

8. $8 \times 5 =$ _____

$40 \div 8 =$ _____

$45 \div 5 =$ _____

$5 \times 8 =$ _____

$40 \div 5 =$ _____

9. $36 \div 9 =$ _____

$4 \times 9 =$ _____

$9 \times 4 =$ _____

$36 \div 4 =$ _____

$32 \div 4 =$ _____

Write each family of facts using the given numbers.

10. 3, 6, 18

_____

_____

_____

_____

11. 5, 6, 30

_____

_____

_____

_____

12. 4, 3, 12

_____

_____

_____

_____

SHARPEN
YOUR
SKILLS

Dividing by Six

Write each quotient.

1. $24 \div 6 =$ _____ **2.** $36 \div 6 =$ _____ **3.** $30 \div 6 =$ _____

4. $18 \div 6 =$ _____ **5.** $27 \div 3 =$ _____ **6.** $42 \div 6 =$ _____

7. $25 \div 5 =$ _____ **8.** $12 \div 6 =$ _____ **9.** $48 \div 6 =$ _____

10. $40 \div 5 =$ _____ **11.** $54 \div 6 =$ _____ **12.** $6 \div 3 =$ _____

13. $6\overline{)18}$ **14.** $6\overline{)48}$ **15.** $4\overline{)28}$ **16.** $6\overline{)30}$

17. $5\overline{)15}$ **18.** $6\overline{)36}$ **19.** $6\overline{)12}$ **20.** $4\overline{)32}$

21. $6\overline{)54}$ **22.** $3\overline{)15}$ **23.** $6\overline{)24}$ **24.** $6\overline{)42}$

Solve each problem.

25. At the bakery, Susann helps fill boxes with crackers. Susann had 54 crackers to divide evenly into 6 boxes. How many crackers fit in each box?

26. Susann put 6 loaves of bread onto each tray. She had 48 loaves of bread. How many trays did she fill?

Dividing by Seven

What animal never leaves its house?

To answer the question, first write each quotient.
Then connect the dots in the order of the answers.

1. $56 \div 7 = \underline{\quad 8 \quad}$ **2.** $36 \div 6 = \underline{\quad 6 \quad}$ **3.** $35 \div 7 = \underline{\quad\quad}$

4. $18 \div 6 = \underline{\quad\quad}$ **5.** $49 \div 7 = \underline{\quad\quad}$ **6.** $14 \div 7 = \underline{\quad\quad}$

7. $63 \div 7 = \underline{\quad\quad}$ **8.** $24 \div 6 = \underline{\quad\quad}$ **9.** $32 \div 4 = \underline{\quad\quad}$

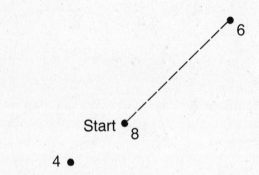

Home
Sweet
Home

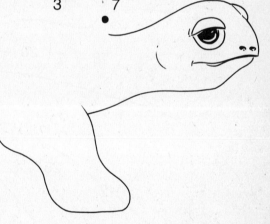

Write each quotient.

10. $21 \div 7 = \underline{\quad\quad}$ **11.** $48 \div 6 = \underline{\quad\quad}$ **12.** $28 \div 7 = \underline{\quad\quad}$

13. $32 \div 4 = \underline{\quad\quad}$ **14.** $42 \div 7 = \underline{\quad\quad}$ **15.** $35 \div 5 = \underline{\quad\quad}$

16. $6\overline{)42}$ **17.** $7\overline{)28}$ **18.** $4\overline{)28}$ **19.** $3\overline{)21}$

20. $3\overline{)12}$ **21.** $5\overline{)30}$ **22.** $3\overline{)27}$ **23.** $2\overline{)14}$

Dividing by Eight

Divide. Color the shape for each answer.

Answer	3	4	5	6	7
Color	Orange	Blue	Red	Brown	Yellow

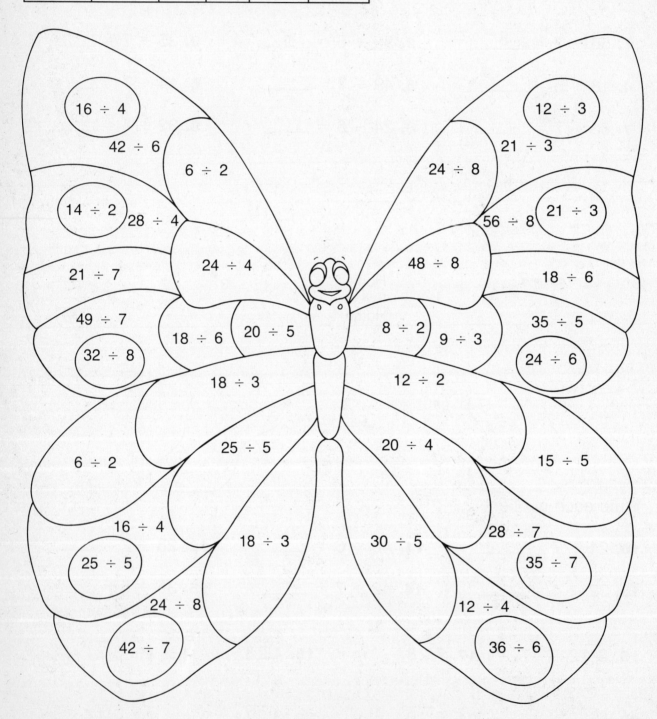

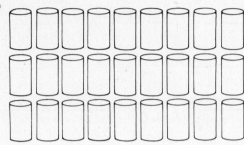

Dividing by Nine

Write each quotient.

1.

$27 \div 9 = \underline{\hspace{1cm}}$

2.

$45 \div 9 = \underline{\hspace{1cm}}$

3. $54 \div 9 = \underline{\hspace{1cm}}$ **4.** $63 \div 9 = \underline{\hspace{1cm}}$ **5.** $27 \div 9 = \underline{\hspace{1cm}}$

6. $45 \div 9 = \underline{\hspace{1cm}}$ **7.** $18 \div 9 = \underline{\hspace{1cm}}$ **8.** $72 \div 9 = \underline{\hspace{1cm}}$

9. $81 \div 9 = \underline{\hspace{1cm}}$ **10.** $36 \div 9 = \underline{\hspace{1cm}}$ **11.** $42 \div 6 = \underline{\hspace{1cm}}$

12. $4\overline{)28}$ **13.** $9\overline{)54}$ **14.** $7\overline{)56}$ **15.** $9\overline{)81}$

16. $9\overline{)63}$ **17.** $8\overline{)56}$ **18.** $9\overline{)18}$ **19.** $7\overline{)42}$

Solve each problem.

20. Sandy collected 72 cans in 9 days. She collected the same number each day. How many cans did she collect each day?

21. Sandy put 45 of the cans in rows. There were 9 cans in each row. How many rows did she have?

Families of Facts

Write each answer. Write number sentences
to complete each family of facts.

1. $5 \times 9 =$ _____

_____

_____

_____

2. $7 \times 8 =$ _____

_____

_____

_____

3. $6 \times 5 =$ _____

_____

_____

_____

Write the family of facts for each picture.

4.

_____

_____

_____

5.

_____

_____

_____

6.

_____

_____

_____

Write each answer. Circle the fact that does not belong to the family.

7. $6 \times 3 =$ _____

$3 \times 6 =$ _____

$18 \div 3 =$ _____

$6 \times 6 =$ _____

$18 \div 6 =$ _____

8. $32 \div 4 =$ _____

$7 \times 4 =$ _____

$28 \div 4 =$ _____

$4 \times 7 =$ _____

$28 \div 7 =$ _____

9. $54 \div 9 =$ _____

$6 \times 9 =$ _____

$45 \div 9 =$ _____

$9 \times 6 =$ _____

$54 \div 6 =$ _____

Choose an Operation

Write whether you add, subtract, multiply, or divide.
Then find each answer.

1. A flower shop had 42 tulip bulbs. The bulbs were packaged 6 to a box. How many boxes were there?

_____

2. The shop had 324 bunches of violets on Saturday morning. That day 178 bunches were sold. How many bunches of violets were left?

_____

3. There were 36 lilies used in 4 arrangements. The same number of lilies was used in each. How many lilies were used in each arrangement?

_____

4. There were 27 daffodils in one vase and 36 in another vase. How many daffodils were there in all in the two vases?

_____

5. One week 246 carnations arrived. The next week 327 arrived. How many carnations arrived during those two weeks?

_____

6. Each rose bush had 5 buds. How many buds were on 8 rose bushes?

_____

7. Marigold seeds were in packages of 6. How many marigold seeds were in 9 packages?

_____

8. The shop keeps 95 plants in the store. There are 14 that need water each day. How many do not need water each day?

_____

Zero and One in Division

Write each quotient.

1. $3 \div 3 =$ _____

2. $4 \div 1 =$ _____

3. $0 \div 5 =$ _____

4. $6 \div 6 =$ _____

5. $7 \div 1 =$ _____

6. $0 \div 8 =$ _____

7. $9 \div 9 =$ _____

8. $0 \div 1 =$ _____

9. $2 \div 1 =$ _____

10. $0 \div 3 =$ _____

11. $4 \div 4 =$ _____

12. $5 \div 1 =$ _____

13. $0 \div 6 =$ _____

14. $7 \div 7 =$ _____

15. $8 \div 1 =$ _____

16. $0 \div 9 =$ _____

17. $1 \div 1 =$ _____

18. $2 \div 2 =$ _____

19. $3 \div 1 =$ _____

20. $0 \div 4 =$ _____

21. $6 \div 1 =$ _____

22. $5 \div 5 =$ _____

23. $0 \div 7 =$ _____

24. $8 \div 8 =$ _____

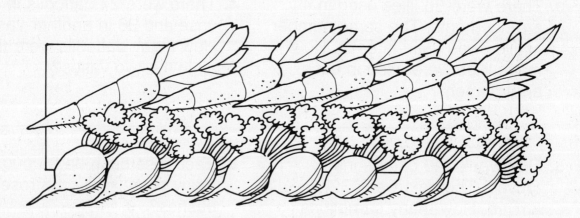

Solve each problem.

25. Ellen grew 8 carrots. She gave the same number of carrots to each of 8 friends. How many carrots did she give each friend?

26. Curtis had 6 radishes. He packaged them in groups of 6. How many packages did he make?

_____

_____

Use after pages 344–345.

Remainders

The Monster of the Deep has trapped Diver Dan.
Help Diver Dan escape to safety.

Use small objects to help you answer
each exercise. **Remember** to tell the
remainder. Then follow your answers
in order through the maze.

1. $3\overline{)14}$ **2.** $5\overline{)19}$ **3.** $4\overline{)27}$ **4.** $7\overline{)15}$

5. $9\overline{)30}$ **6.** $4\overline{)26}$ **7.** $3\overline{)24}$ **8.** $4\overline{)31}$

9. $4\overline{)34}$ **10.** $7\overline{)17}$ **11.** $2\overline{)13}$ **12.** $3\overline{)23}$

SAFETY

7 R2	3 R3	2 R1	6 R3	5 R4
6 R1	6 R2	4 R2	3 R4	6 R4
2 R3	8		4	3 R3
8 R2	7 R3	4 R1	5 R2	7 R1

Use after pages 346–347.

Interpreting Remainders

Draw a picture to help you answer each exercise.

1. Ed is chopping wood for firewood. He has 30 pieces of wood. He can fit 7 pieces in his wheelbarrow. How many wheelbarrow loads will it take to move the wood?

2. How many pieces of wood will be in the last load in Problem 1?

_____

3. Ed's truck holds 60 pieces of wood. He is selling wood to 8 people. Each person is getting the same amount of wood. How many pieces will be left over?

4. Leonard is setting the tables in the lunch room. He has 75 forks. If 8 people can sit at a table, how many full tables can he set with forks?

_____

5. How many forks will be left over in Problem 4?

6. Each table in the dining room has 3 pepper shakers on it. If Leonard has 29 pepper shakers, how many tables can he set?

_____

Fractions: Part of a Whole

Circle the fraction that tells how much is shaded.

1.

$\frac{2}{3}$ $\frac{1}{4}$ $\frac{3}{4}$

2.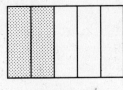

$\frac{2}{5}$ $\frac{3}{6}$ $\frac{2}{6}$

3.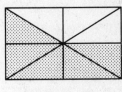

$\frac{6}{7}$ $\frac{5}{8}$ $\frac{5}{6}$

4.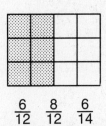

$\frac{6}{12}$ $\frac{8}{12}$ $\frac{6}{14}$

5.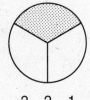

$\frac{2}{3}$ $\frac{2}{4}$ $\frac{1}{3}$

6.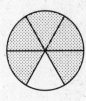

$\frac{5}{7}$ $\frac{5}{6}$ $\frac{5}{8}$

Write the fraction that tells how much is
shaded in each picture.

7.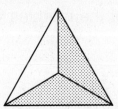

_____

8.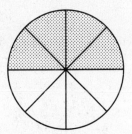

_____

9.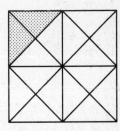

_____

10.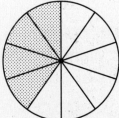

_____

11.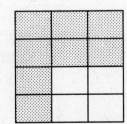

_____

12.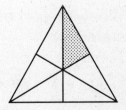

_____

Use after pages 360–363.

NAME

Fractions: Part of a Set

Write each fraction.

1. What fraction of the toys are miniature cars?

 _____

2. What fraction of the trees have leaves?

 _____

3. What fraction of the frogs are on the lily pad?

 _____

4. What fraction of the pencils are not sharpened?

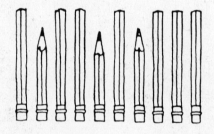 _____

5. What fraction of the darts are on the dart board?

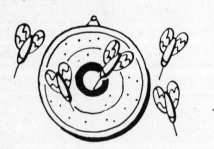 _____

6. What fraction of the marbles have stripes?

 _____

7. What fraction of the apples have stems?

 _____

8. Of the 8 members of the gym club, 5 are girls. What fraction of the club members are girls?

 _____

SHARPEN
YOUR
SKILLS

Equal Fractions

Complete each number sentence to show equal fractions.

1.
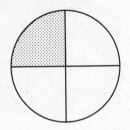

$$\frac{1}{4} = \frac{}{8}$$

2.
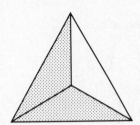

$$\frac{2}{3} = \frac{}{6}$$

3.
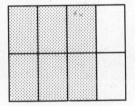

$$\frac{6}{8} = \frac{}{}$$

4.
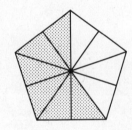

$$\frac{6}{10} = \frac{}{}$$

5.
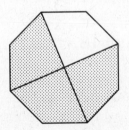

$$\frac{}{} = \frac{}{}$$

6.
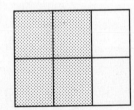

$$\frac{}{} = \frac{}{}$$

7. Name two fractions that are equal to $\frac{3}{4}$.

_____

8. Name two fractions that are equal to $\frac{1}{3}$.

_____

SHARPEN YOUR SKILLS

Finding Fractional Parts

Find each answer. Make a picture if you need to.

1.

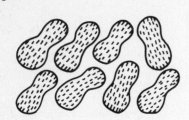

$\frac{1}{4}$ of 8 = _____

2.

$\frac{1}{2}$ of 6 = _____

3.

$\frac{1}{3}$ of 9 = _____

4.

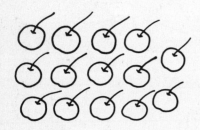

$\frac{1}{2}$ of 14 = _____

5.

$\frac{1}{3}$ of 12 = _____

6.

$\frac{1}{5}$ of 10 = _____

7. $\frac{1}{5}$ of 20 = _____

8. $\frac{1}{6}$ of 24 = _____

9. $\frac{1}{6}$ of 18 = _____

10. $\frac{1}{5}$ of 25 = _____

11. $\frac{1}{2}$ of 30 = _____

12. $\frac{1}{5}$ of 30 = _____

13. $\frac{1}{2}$ of 28 = _____

14. $\frac{1}{5}$ of 45 = _____

15. $\frac{1}{6}$ of 48 = _____

Solve each problem.

16. Mark picked $\frac{1}{3}$ of the 12 pumpkins in his garden. How many pumpkins did he pick?

17. Betty had 15 oranges. She put 5 oranges in each sack. How many sacks did she use?

_____   _____

Practice/EXPLORING MATHEMATICS © Scott, Foresman and Company/3

Try and Check

Solve each problem.

1. At a rodeo there are horses and people. In all, there are 48 legs. There are twice as many people as horses. How many horses are there?

2. How many people are there in Problem 1?

3. The rodeo charges $4 for adults and $3 for children. Jaime gives the cashier $23 for 7 people. How many children are there?

4. How many adults are there in Problem 3?

5. Phillip is looking through a telescope at the sky. He sees 46 stars in all in 7 constellations. Each constellation has either 6 or 7 stars. How many have 6 stars?

6. How many constellations in Problem 5 have 7 stars?

Comparing Fractions

Complete each number sentence using >, <, or =.

1.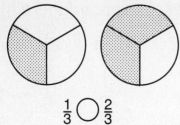

$\frac{1}{3} \bigcirc \frac{2}{3}$

2.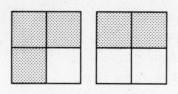

$\frac{3}{4} \bigcirc \frac{2}{4}$

3.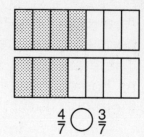

$\frac{4}{7} \bigcirc \frac{3}{7}$

4.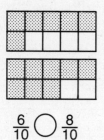

$\frac{6}{10} \bigcirc \frac{8}{10}$

5.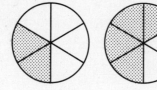

$\frac{2}{6} \bigcirc \frac{4}{6}$

6.

$\frac{5}{8} \bigcirc \frac{5}{8}$

7.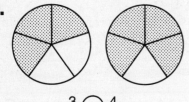

$\frac{3}{5} \bigcirc \frac{4}{5}$

8.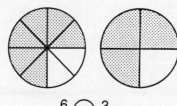

$\frac{6}{8} \bigcirc \frac{3}{4}$

9.

$\frac{2}{3} \bigcirc \frac{1}{3}$

10.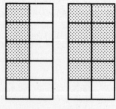

$\frac{4}{10} \bigcirc \frac{8}{10}$

11.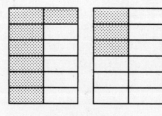

$\frac{7}{12} \bigcirc \frac{3}{12}$

12.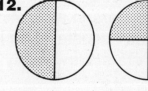

$\frac{1}{2} \bigcirc \frac{2}{4}$

13. $\frac{7}{8}$ _____ $\frac{3}{8}$

14. $\frac{3}{9}$ _____ $\frac{1}{3}$

15. $\frac{1}{6}$ _____ $\frac{1}{5}$

16. $\frac{1}{2}$ _____ $\frac{1}{3}$

17. $\frac{2}{8}$ _____ $\frac{1}{4}$

18. $\frac{1}{4}$ _____ $\frac{2}{3}$

Ordering Fractions

What did one math book say to the other math book?

Write the fractions in order from least to greatest. Use the fraction strips below to help. To answer the question, find the matching group of fractions at the bottom of the page and write the letter for that exercise.

1. $\frac{5}{8}$ $\frac{4}{6}$ $\frac{4}{5}$

_____ M

2. $\frac{3}{6}$ $\frac{2}{5}$ $\frac{3}{8}$

_____ R

3. $\frac{4}{10}$ $\frac{1}{4}$ $\frac{2}{3}$

_____ P

4. $\frac{1}{6}$ $\frac{2}{8}$ $\frac{3}{5}$

_____ O

5. $\frac{3}{3}$ $\frac{3}{4}$ $\frac{5}{8}$

_____ !

6. $\frac{9}{10}$ $\frac{8}{8}$ $\frac{5}{6}$

_____ B

7. $\frac{5}{6}$ $\frac{7}{8}$ $\frac{4}{5}$

_____ S

8. $\frac{2}{5}$ $\frac{3}{8}$ $\frac{5}{10}$

_____ L

9. $\frac{3}{4}$ $\frac{6}{10}$ $\frac{2}{3}$

_____ E

$\frac{1}{2}$		$\frac{1}{2}$	
$\frac{1}{3}$	$\frac{1}{3}$	$\frac{1}{3}$	
$\frac{1}{4}$	$\frac{1}{4}$	$\frac{1}{4}$	$\frac{1}{4}$

$\frac{1}{5}$ $\frac{1}{5}$ $\frac{1}{5}$ $\frac{1}{5}$ $\frac{1}{5}$

$\frac{1}{6}$ $\frac{1}{6}$ $\frac{1}{6}$ $\frac{1}{6}$ $\frac{1}{6}$ $\frac{1}{6}$

$\frac{1}{8}$ $\frac{1}{8}$ $\frac{1}{8}$ $\frac{1}{8}$ $\frac{1}{8}$ $\frac{1}{8}$ $\frac{1}{8}$ $\frac{1}{8}$

$\frac{1}{10}$ $\frac{1}{10}$ $\frac{1}{10}$ $\frac{1}{10}$ $\frac{1}{10}$ $\frac{1}{10}$ $\frac{1}{10}$ $\frac{1}{10}$ $\frac{1}{10}$ $\frac{1}{10}$

"I have ___ ___ ___ ___ ___ ___ ___ ___ ___"

$\frac{1}{4}, \frac{4}{10}, \frac{2}{3}$ $\frac{3}{8}, \frac{2}{5}, \frac{3}{6}$ $\frac{1}{6}, \frac{2}{8}, \frac{3}{5}$ $\frac{5}{6}, \frac{9}{10}, \frac{8}{8}$ $\frac{3}{8}, \frac{2}{5}, \frac{5}{10}$ $\frac{6}{10}, \frac{2}{3}, \frac{3}{4}$ $\frac{5}{8}, \frac{4}{6}, \frac{4}{5}$ $\frac{4}{5}, \frac{5}{6}, \frac{7}{8}$ $\frac{5}{8}, \frac{3}{4}, \frac{3}{3}$

Tenths

Write the decimal and the fraction for the shaded parts.

1. _____ **2.** _____ **3.** _____

Write the decimal.

4. Eight tenths _____ **5.** Two tenths _____

6. $\frac{5}{10}$ _____ **7.** $\frac{3}{10}$ _____ **8.** $\frac{7}{10}$ _____

Write the fraction.

9. 0.7 _____ **10.** 0.6 _____

11. 0.2 _____ **12.** 0.5 _____

13. 0.4 _____ **14.** 0.9 _____

Solve each problem.

Carl had a full container of lemonade for his stand. The picture at the right shows the amount of lemonade he has left.

15. How many tenths of a container of lemonade did he sell?

_____

16. How many tenths of a container of lemonade are left?

_____

Decimal Place Value

Underline the tenths digit in each exercise.

1. 34.6　　　　**2.** 29.1　　　　**3.** 58.7　　　　**4.** 32.9　　　　**5.** 35.2

Underline the tens digit in each exercise.

6. 12.7　　　　**7.** 38.6　　　　**8.** 45.4　　　　**9.** 56.1　　　　**10.** 11.3

Tell what place the 6 is in for each exercise.

11. 63.3　　　　**12.** 56.3　　　　**13.** 36.7　　　　**14.** 23.6　　　　**15.** 98.6

_____　　_____　　_____　　_____　　_____

Write each number as a decimal.

16. twenty-seven and nine tenths

_____

17. eleven and three tenths

_____

18. nineteen and six tenths

_____

19. thirty-four and two tenths

_____

20. forty-five and three tenths

_____

21. fifty-six and eight tenths

_____

22. seventy and nine tenths

_____

23. five and one tenth

_____

24. ninety-two and four tenths

_____

25. sixty-six and six tenths

_____

SHARPEN
YOUR
SKILLS

Hundredths

Write each decimal.

1.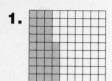

2.

3.

_____   _____   _____

4. Six hundredths _____   **5.** Fifteen hundredths _____

6. Nine and sixty-six hundredths _____

7. Three and fifty-one hundredths _____

8. Seventy-eight hundredths _____

9. Thirty-one hundredths _____

10. $\frac{75}{100}$ = _____   **11.** $\frac{29}{100}$ = _____   **12.** $\frac{7}{100}$ = _____   **13.** $\frac{68}{100}$ = _____

Tell whether each pair names the same
number. Write Yes or No.

14. $\frac{25}{100}$ 2.50   **15.** $4\frac{75}{100}$ 4.75   **16.** $\frac{54}{100}$ 0.54

_____   _____   _____

17. $12\frac{12}{100}$ 1.12   **18.** $\frac{34}{100}$ 3.04   **19.** $4\frac{5}{100}$ 4.05

_____   _____   _____

20. $3\frac{95}{100}$ 39.5   **21.** $\frac{9}{100}$ 0.09   **22.** $44\frac{65}{100}$ 44.65

_____   _____   _____

Comparing Decimals

Write each number sentence using < or >.

1. 3.45 _____ 2.99

2. 0.7 _____ 0.79

3. 1.01 _____ 1.10

4. 2.4 _____ 2.04

5. 0.65 _____ 0.63

6. 2.54 _____ 1.87

Shade in the decimal models for both
numbers in each exercise. Then write the
number sentence using < or >.

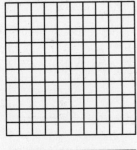
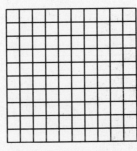
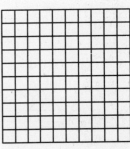

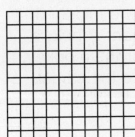
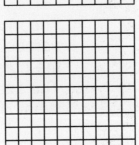
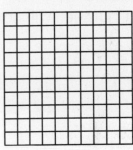

7. 0.54 _____ 0.63

8. 1.0 _____ 0.99

9. 0.09 _____ 0.11

Write two number sentences using > and <
for each pair of numbers.

10. 3.98, 3.9

11. 4.75, 3.99

12. 2.01, 2.5

_____

_____

_____

_____

_____

_____

Use Alternate Strategies

Think of two strategies you could use to solve each
problem. Use both ways to find the answer.

1. Kerry buys 10 peaches. She puts
0.4 of them on the table. She
puts one tenth of them in the
refrigerator. How many are left?

2. Kerry eats one of the peaches.
How many are left in all? Write
the answer as a decimal.

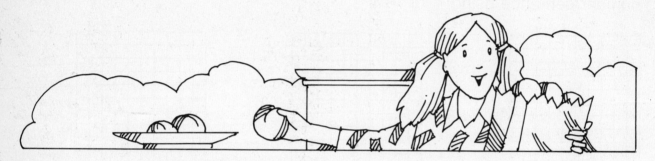

3. What number will come next?

3, 6, 4, 7, 5, 8 _____

4. The monkey is in the middle of a
parade. The elephant is before
the horse and after the monkey.
The zebra is before the camel.
What is the order of animals?

5. Jerri is putting 100 candles on
her great-grandmother's birthday
cake. One half the candles are
on the cake. Jerri has 0.24 of the
candles in her hand. How many
are still in the box? What part of
the candles is this?

6. Barry rode a total of 10 miles on
his bike on Saturday. He reached
Mill Creek at noon and had
ridden half the distance. At
3 o'clock he had 1 mile left to go.
How far did he ride between
noon and 3:00?

Using Decimals

About how long is each object?

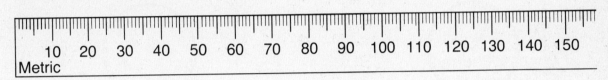

1.

_____

2.

_____

3.

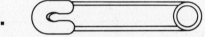

_____

4.

_____

Write how much gas will be used when another $\frac{2}{10}$ of a gallon are pumped.

5.

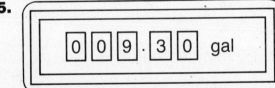

_____

6.

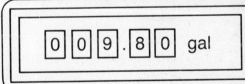

_____

7.

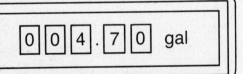

_____

8.

_____

**SHARPEN
YOUR
SKILLS**

Minutes Before the Hour

Write each time as minutes before the hour.

1.

_____

2.

_____

3.

_____

4.

_____

Show the time by drawing the hands on each clockface below.

5. 22 minutes before 3

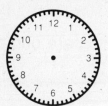

6. 9 minutes before 1

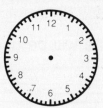

7. 16 minutes after 4

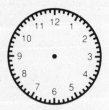

8. 20 minutes before 8

Problem Solving   Write the times below as minutes before the hour.

9. Larry wished to arrive at the movies at 20 minutes before 3:00. He was 5 minutes early. What time was it when he arrived?

_____

10. Cindy arrived at the movies 10 minutes after Larry. What time was it when she arrived?

_____

Practice/EXPLORING MATHEMATICS © Scott, Foresman and Company/3

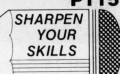

SHARPEN
YOUR
SKILLS

Calendar: Year

June							July							August						
S	M	T	W	T	F	S	S	M	T	W	T	F	S	S	M	T	W	T	F	S
					1	2	1	2	3	4	5	6	7				1	2	3	4
3	4	5	6	7	8	9	8	9	10	11	12	13	14	5	6	7	8	9	10	11
10	11	12	13	14	15	16	15	16	17	18	19	20	21	12	13	14	15	16	17	18
17	18	19	20	21	22	23	22	23	24	25	26	27	28	19	20	21	22	23	24	25
24	25	26	27	28	29	30	29	30	31					26	27	28	29	30	31	

Use the calendar to answer each question.

1. What day of the week is July 5?

2. What day of the week is June 12?

_____

_____

3. What day of the week is August 15?

4. What day of the week is July 16?

_____

_____

Write the date.

5. the third Monday in June

6. the second Saturday in July

_____

_____

7. the fifth Monday in July

8. the first Tuesday in August

_____

_____

Solve each problem.

9. How many weeks are there between July 4 and August 22?

10. Which months have 31 days?

_____

_____

Multiple-Step Problems

```
TRAIN SCHEDULE FROM CENTERVILLE
        Leave Centerville 12:30
```

Arrive	Time
Grayson	1:05
Littleton	1:15
Smithtown	1:25
Freesburg	1:40
Green Valley	1:55
Watertown	2:10

Use the schedule to answer each question.

1. Mary Lou rides the train from Centerville to Freesburg and then walks home. The walk takes her 15 minutes. At what time does she arrive home?

2. Buddy rides the train from Centerville to Grayson. He then rides his bicycle for 8 minutes to get home. At what time does he arrive home? How long is the total trip?

3. Shirley walks for 20 minutes from her job to the train in Centerville. Then she takes the train to Watertown. How long does it take her to get to Watertown?

4. It takes Ben 7 minutes to reach the station in Centerville. At what time must he leave to catch the train?

How long will it take him to go from his home to Smithtown?

Critical Thinking   Jason wants to ride the train from Freesburg to Centerville. If he leaves at 9:05 in the morning, at what time will he arrive in Centerville? The train takes the same amount of time to travel in either direction.

Identifying Coins and Bills

Write the letters of two items in the second column that
match each coin or bill in the first column.

1. _____
   _____

2. _____
   _____

3. _____
   _____

4. _____
   _____

5. _____
   _____

6. _____
   _____

7. _____
   _____

8. _____
   _____

a. one dollar
b. $10.00

c. 5 cents
d. $0.10

e. half-dollar
f. 25 cents

g. 1¢
h. $0.05

i. ten dollars
j. $0.50

k. $0.25
l. 10¢

m. $0.01
n. $1.00

o. $5.00
p. five dollars

Write the name of the coin or bill.

9. I am worth more than a penny,
   but less than a dime.

10. I am worth more than a nickel,
    but less than a quarter.

_____

_____

SHARPEN YOUR SKILLS

Counting Money

Use play money to complete Exercises 1–13.
For Exercises 1–6, write each amount in words.

1. $6.33

_____

_____

2. $5.27

_____

_____

3. $9.54

_____

_____

4. $2.87

_____

_____

5. $4.04

_____

_____

6. $1.46

_____

_____

For Exercises 7–13, write the amount of money.
Remember to use a $ and . in your answers.

7. 3 dimes,
2 quarters

_____

8. 2 quarters,
1 dime,
3 pennies

_____

9. 1 nickel,
8 pennies

_____

10. 2 dimes,
4 nickels,
6 pennies

_____

11. 4 dollars,
2 quarters, 1 dime,
3 nickels, 1 penny

_____

12. 1 dollar,
1 half-dollar,
1 quarter, 1 dime,
2 nickels,
4 pennies

_____

13. 3 dollars, 1 quarter,
3 dimes, 2 pennies

_____

Adding and Subtracting Money

Add or subtract. **Remember** to line up the decimal points and place $ and . in your answers.

1. $19.95
    +   1.76
    _____

2. $3.06
    + 3.52
    _____

3. $14.37
    − 9.23
    _____

4. $2.98
    − 1.45
    _____

5. $30.25
    − 10.53
    _____

6. $23.67
    + 18.68
    _____

7. $4.54
    − 3.22
    _____

8. $6.72
    + 5.09
    _____

9. $50.55
    − 18.03
    _____

10. $5.97
    − 2.65
    _____

11. $38.19
    + 36.12
    _____

12. $5.42
    + 4.97
    _____

Solve each problem.

13. Renee bought a pair of shoes for $9.99 and a hat for $3.47. How much did she spend in all?

14. Jaime bought a shirt for $5.55 and three pairs of socks for $3.33. How much less did he spend than Renee did?

15. Liza bought an aquarium for $7.95 and a turtle for $4.75. How much did she spend in all?

16. Toby bought a puppy for $32.50 and a dog basket for $6.05. How much more did he spend than Liza did?

Making Change

Write the amount of change. Can you do
some of these problems mentally?

1.

Paid $5.00

_____

2.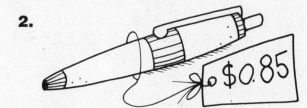

Paid $1.00

_____

3.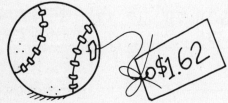

Paid $2.00

_____

4.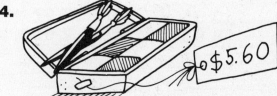

Paid $10.00

_____

5.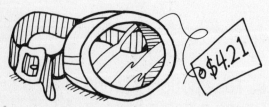

Paid $5.00

_____

6.

Paid $2.50

_____

Work Backward

Choose mental math, paper and pencil, or a calculator to solve each problem.

1. Naomi planted 14 pepper plants, some potatoes, and 12 tomato plants. Altogether, she planted 36 plants. How many potatoes did she plant?

_____

2. Louise picked 75 peppers, carrots, and squash. She picked 30 squash and 28 carrots. How many peppers did she pick?

_____

3. Hiro planted 6 cucumbers, 30 carrots, and some lettuce plants in his garden. Altogether, he planted 50 plants. How many lettuce plants did he plant?

_____

4. Bruce picked 35 tomatoes last week. This week he has picked 42. There are 16 tomatoes left in the garden. How many tomatoes were there before he picked them?

_____

5. Ginger spent $12 on plants, $4 on fertilizer, and $6 on materials for a fence. She had $8 left. How much money did she have at first?

_____

6. Norma weeded her garden for 25 minutes and turned the soil for 15 minutes. Then she rested for 10 minutes and left at 11:30. What time did she begin?

_____

7. Avram wants to do his math assignment and write his book report before dinner at 6.00 P.M. The math takes 30 minutes and the book report takes $1\frac{1}{2}$ hours. What time should he begin? _____

Critical Thinking   Luís wants to do his homework before his violin lesson. The lesson is at the teacher's house, and it starts at 4:00 P.M. What else must you know to figure out what time he should begin?

_____

_____

Estimation and Money

Estimate each amount by rounding to the
nearest dollar. Write the estimated sum.
Then use a calculator to find the actual sum.

1. $2.79 + $0.32 + $4.63

2. $0.96 + $3.64 + $4.19

_____

3. $14.34 + $9.99 + $13.06

4. $12.62 + $3.98 + $15.75

_____

5. $3.89 + $6.78 + $1.12

6. $33.15 + $5.98 + $2.09

_____

7. $4.44 + $0.89 + $3.02

8. $12.10 + $4.97 + $7.75
+ $15.15

_____

9. $54.43 + $22.67 + $3.11

10. $3.98 + $63.09 + $23.19
+ $44.78

_____

11. Which estimate is the closest to
the exact answer?

12. Which estimate is the farthest
from the exact answer?

_____

Exploring Multiplication

Give the standard form for each exercise.
Use place-value materials to help you.

1. 3 hundreds

_____

2. 15 tens

_____

3. 6 hundreds

_____

4. 18 hundreds

_____

5. 8 tens

_____

6. 12 tens

_____

7. 3 thirties

_____

8. 2 twenties

_____

9. 2 eighties

_____

10. 4 groups of two hundreds

_____

11. 6 groups of five hundreds

_____

12. 8 groups of one hundred

_____

13. 3 groups of two hundreds

_____

Problem Solving Solve each problem.

14. Jason bought 10 bags of
peanuts. Each bag had
24 peanuts in it. How many
peanuts did he buy in all?

_____

15. Penelope bought 6 bags of
nails. Each bag had 30 nails in
it. How many nails did she buy
in all?

_____

SHARPEN YOUR SKILLS

Multiples of 10 and 100: Mental Math

Mental Math Find each product mentally.

1. $4 \times 60 =$ _____   **2.** $30 \times 7 =$ _____   **3.** $40 \times 8 =$ _____

4. $8 \times 80 =$ _____   **5.** $20 \times 3 =$ _____   **6.** $5 \times 30 =$ _____

7. $200 \times 6 =$ _____   **8.** $9 \times 300 =$ _____   **9.** $7 \times 400 =$ _____

10. $900 \times 6 =$ _____   **11.** $4 \times 400 =$ _____   **12.** $500 \times 5 =$ _____

13. $3 \times 30 =$ _____   **14.** $4 \times 90 =$ _____   **15.** $8 \times 700 =$ _____

16. $70 \times 2 =$ _____   **17.** $600 \times 8 =$ _____   **18.** $2 \times 20 =$ _____

19. $500 \times 4 =$ _____   **20.** $7 \times 700 =$ _____   **21.** $700 \times 3 =$ _____

22. $600 \times 3 =$ _____   **23.** $2 \times 90 =$ _____   **24.** $80 \times 9 =$ _____

Complete each table.

25.

×	2	9	7	5	8	3
50	100					

26.

×	4	3	6	9	7	2
800						

Estimating Products

Estimate each product.

1. 5×22 **2.** 3×48 **3.** 71×4 **4.** 7×57

_____   _____   _____   _____

5. 3×221 **6.** 578×4 **7.** 7×489 **8.** 6×324

_____   _____   _____   _____

9. 5×382 **10.** 9×54 **11.** 563×4 **12.** 128×4

_____   _____   _____   _____

13. 43×9 **14.** 8×546 **15.** 3×333 **16.** 972×6

_____   _____   _____   _____

Critical Thinking Solve these problems.

17. In which of Exercises 1–16 do you think that your estimated product will be less than the actual product?

_____

_____

18. In which of Exercises 1–16 do you think that your estimated product will be greater than the actual product?

_____

_____

Exploring Multiplication Further

Use place-value materials to help you
find how many are in

1. 3 groups of 16.

_____

_____

2. 6 rows of 12.

_____

_____

3. 5 groups of 21.

_____

_____

4. 4 rows of 18.

_____

_____

5. 5 groups of 42.

_____

_____

6. 3 rows of 55.

_____

_____

7. 6 groups of 15.

_____

_____

8. 4 groups of 26.

_____

_____

9. 3 rows of 36.

_____

_____

10. 3 rows of 54.

_____

_____

11. 6 groups of 17.

_____

_____

12. 5 rows of 44.

_____

_____

Problem Solving Solve each problem.

13. At a swimming meet there are 8
races. Each race has a total of
16 people racing. How many
racers are there in all?

_____

14. A baseball league has 8 teams.
Each team has 15 players on it.
How many players are in the
league?

_____

Multiplication: Renaming Ones

The more you crack it, the more you will be liked. What is it?

To find out, multiply. Then connect
the dots in order as you go.

1. $\begin{array}{r} 28 \\ \times\ 3 \\ \hline 84 \end{array}$
 2. $\begin{array}{r} 39 \\ \times\ 2 \\ \hline 78 \end{array}$
 3. $\begin{array}{r} 16 \\ \times\ 2 \\ \hline \end{array}$
 4. $\begin{array}{r} 13 \\ \times\ 4 \\ \hline \end{array}$

5. $\begin{array}{r} 13 \\ \times\ 5 \\ \hline \end{array}$
 6. $\begin{array}{r} 16 \\ \times\ 3 \\ \hline \end{array}$
 7. $\begin{array}{r} 25 \\ \times\ 3 \\ \hline \end{array}$
 8. $\begin{array}{r} 16 \\ \times\ 6 \\ \hline \end{array}$

9. $\begin{array}{r} 13 \\ \times\ 7 \\ \hline \end{array}$
 10. $\begin{array}{r} 14 \\ \times\ 5 \\ \hline \end{array}$
 11. $\begin{array}{r} 17 \\ \times\ 4 \\ \hline \end{array}$
 12. $\begin{array}{r} 19 \\ \times\ 3 \\ \hline \end{array}$

13. $2 \times 17 =$ _____

14. $6 \times 15 =$ _____

15. $2 \times 18 =$ _____

16. $2 \times 29 =$ _____

17. $3 \times 24 =$ _____

18. $3 \times 17 =$ _____

19. $3 \times 14 =$ _____

20. $3 \times 27 =$ _____

21. $7 \times 12 =$ _____

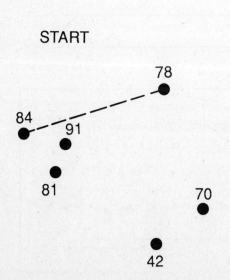
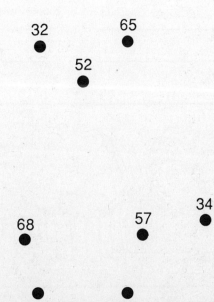
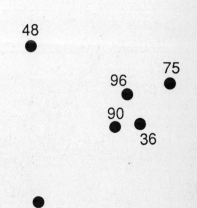

SHARPEN
YOUR
SKILLS

Too Much Information

Solve each problem. Then tell what
information is given that you do not need to know.

1. Last fall, Bob planted 4 shrubs
and 13 rows of trees with 8 trees
in each row. How many trees did
he plant in all?

_____

2. Amy planted 6 tree seedlings on
her family's tree farm. Amy's
parents planted 27 times as
many tree seedlings as she did
and 63 shrubs. How many tree
seedlings did her parents plant?

_____

3. Clay bought 6 potted plants for
59¢ each. Wilma spent 75¢ for a
potted plant. How much did Clay
spend in all?

_____

4. Sheila spent $42.95 for trees and
$7.68 for a shovel. Bob spent
$1.29 for plant food. How much
did Sheila spend in all?

_____

5. Tom planted 34 rows of trees
and 274 shrubs at the tree farm.
There were 9 trees planted in
each row. How many trees did
Tom plant in all?

_____

6. The tree farm sold 6 apple trees
and 4 pear trees each day for 14
days. How many apple trees in
all were sold during that time?

_____

SHARPEN
YOUR
SKILLS

Multiplication: Renaming Ones and Tens

Where does a witch keep her spaceship?

To find out, multiply. **Remember** to estimate first. Each time an answer is given in the puzzle below, write the letter for that exercise. Two answers are not used in the puzzle.

1. $\begin{array}{r} 72 \\ \times\ 6 \\ \hline \end{array}$
N

2. $\begin{array}{r} 54 \\ \times\ 7 \\ \hline \end{array}$
C

3. $\begin{array}{r} 38 \\ \times\ 3 \\ \hline \end{array}$
E

4. $\begin{array}{r} 49 \\ \times\ 7 \\ \hline \end{array}$
R

5. $\begin{array}{r} 73 \\ \times\ 7 \\ \hline \end{array}$
P

6. $\begin{array}{r} 37 \\ \times\ 3 \\ \hline \end{array}$
S

7. $\begin{array}{r} 68 \\ \times\ 3 \\ \hline \end{array}$
O

8. $\begin{array}{r} 93 \\ \times\ 7 \\ \hline \end{array}$
E

9. $\begin{array}{r} 57 \\ \times\ 9 \\ \hline \end{array}$
O

10. $\begin{array}{r} 86 \\ \times\ 5 \\ \hline \end{array}$
M

11. $\begin{array}{r} 25 \\ \times\ 8 \\ \hline \end{array}$
B

12. $\begin{array}{r} 75 \\ \times\ 5 \\ \hline \end{array}$
L

13. $3 \times 69 =$ _____ R

14. $7 \times 43 =$ _____ T

15. $9 \times 89 =$ _____ O

16. $8 \times 34 =$ _____ H

17. $6 \times 85 =$ _____ I

18. $8 \times 27 =$ _____ B

| ___ | ___ | | ___ | ___ | ___ | | ___ | ___ | ___ | ___ | ___ |
| 510 | 432 | | 272 | 651 | 343 | | 216 | 207 | 801 | 513 | 430 |

| ___ | ___ | ___ | | ___ | ___ | ___ |
| 378 | 375 | 204 | | 111 | 114 | 301 |

Choose an Operation

Write whether you would *add, subtract, multiply,* or *divide.*
Then solve each problem.

1. Two families flew across the country. One family flew 2,786 miles. The other flew 2,799 miles. How many miles did the families fly all together?

_____

2. Each person in the 2 families had 3 bags. There were 11 people in all. How many bags were there?

_____

3. Two people flew from New York City to different cities in Europe. Sam flew 5,653 miles. Dave flew 5,322 miles. How much farther did Sam fly than Dave?

_____

4. There were 5 people in one row of an airplane who paid to watch the movie. The total cost was $20. How much did each person pay?

_____

5. Three airplanes flew from San Francisco to Washington, D.C. Flight A took 4 hours and 30 minutes. Flight B took 5 hours. Flight C took 4 hours and 30 minutes. What was the total time of the 3 flights?

_____

6. What is the difference between the fastest and slowest times in Problem 5?

_____

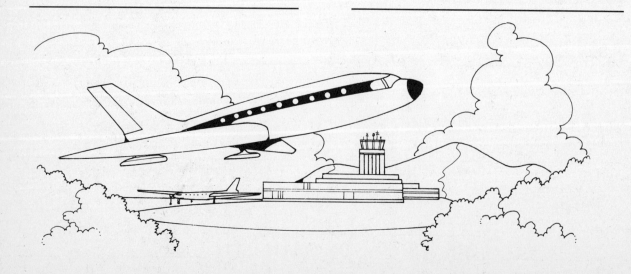

NAME

Multiplication with Hundreds

Find each product. **Remember** to find an estimate first.

1. $\begin{array}{r} 129 \\ \times\ \ 3 \\ \hline \end{array}$

2. $\begin{array}{r} 342 \\ \times\ \ 3 \\ \hline \end{array}$

3. $\begin{array}{r} 219 \\ \times\ \ 5 \\ \hline \end{array}$

4. $\begin{array}{r} 673 \\ \times\ \ 3 \\ \hline \end{array}$

5. $\begin{array}{r} 528 \\ \times\ \ 2 \\ \hline \end{array}$

6. $\begin{array}{r} 432 \\ \times\ \ 7 \\ \hline \end{array}$

7. $\begin{array}{r} 682 \\ \times\ \ 3 \\ \hline \end{array}$

8. $\begin{array}{r} 931 \\ \times\ \ 5 \\ \hline \end{array}$

9. $\begin{array}{r} 241 \\ \times\ \ 9 \\ \hline \end{array}$

10. $\begin{array}{r} 851 \\ \times\ \ 6 \\ \hline \end{array}$

11. $\begin{array}{r} 673 \\ \times\ \ 3 \\ \hline \end{array}$

12. $\begin{array}{r} 527 \\ \times\ \ 3 \\ \hline \end{array}$

13. $\begin{array}{r} 728 \\ \times\ \ 3 \\ \hline \end{array}$

14. $\begin{array}{r} 441 \\ \times\ \ 7 \\ \hline \end{array}$

15. $\begin{array}{r} 213 \\ \times\ \ 5 \\ \hline \end{array}$

16. $\begin{array}{r} 351 \\ \times\ \ 9 \\ \hline \end{array}$

17. $8 \times 512 =$ _____

18. $4 \times 293 =$ _____

19. $3 \times 452 =$ _____

Problem Solving Solve each problem.

20. A farmer sold 392 pints of strawberries each day for 5 days. How many pints of strawberries did he sell in all?

21. Sally made strawberry preserves. Each jar held 245 milliliters. She gave Jeff 6 jars. How many milliliters of preserves is that?

_____

_____

Use after pages 460–461.

Collecting Data

Write two questions you could ask your friends
to help you answer each question.

1. A trip to the zoo was planned for
Tuesday. The class wanted to
make a schedule to decide which
animals to see. Which animals
will students visit?

_____

_____

_____

2. The trip allows 3 hours at the
zoo. What amount of time should
be spent in visiting animals,
eating, and buying souvenirs?

_____

_____

_____

3. There are 3 places to eat at the
zoo. Should the class eat at the
Snack Shop, the Bird Walk, or
the Kitchen Place?

_____

_____

_____

4. Tickets are sold for the water
show. The seats cost $1.00,
$1.25, or $1.50. The class must
choose a price and bring the
money. What price ticket should
the class buy?

_____

_____

_____

5. The class needed to divide into 4
groups for a talk on animal care
and safety. How can the students
be divided into 4 groups of the
same size?

_____

_____

_____

Collecting and Organizing Data

Record data about the information shown.
Complete each tally chart.

1.

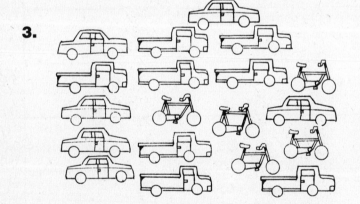

Animals	Tally	Number

2.

Shapes	Tally	Number

3.

Toys	Tally	Number

4.

Clothes	Tally	Number

SHARPEN
YOUR
SKILLS

Pictographs

Complete the pictograph
below by using the tally
chart for the data you
need. Then complete
Exercises 6 and 7.

The tally chart shows
the number of newspapers
collected by the students
in Ms. Anderson's class.

Person	Tally	Number
Gayle	ǁǁ ǁǁ ǁǁ ǁǁ ǁǁ	25
Tom	ǁǁ ǁǁ	10
Lucy	ǁǁ ǁǁ ǁǁ ǁǁ ǁǁ ǁǁ ǁǁ ǁǁ	40
Diane	ǁǁ ǁǁ ǁǁ ǁǁ ǁǁ ǁǁ ǁǁ	35
Joy	ǁǁ ǁǁ ǁǁ ǁǁ	20
Cliff	ǁǁ ǁǁ ǁǁ ǁǁ ǁǁ ǁǁ ǁǁ ǁǁ ǁǁ	45

1. Write a title for your graph.
2. List the persons' names at the left.
3. Use a picture to stand for a newspaper.
4. Decide how many papers each picture will stand for.
5. Draw pictures in your graph.

6. Who collected the most newspapers? _____

7. Describe what your graph shows without using numbers.

_____

Bar Graphs

Use the information on the tally chart to make
a bar graph. Then answer the questions.

Student	Tally	Number
José	ⅢⅡ	7
Maria	ⅢⅢⅠ	11
Ann	Ⅲ	5
Becky	ⅢⅢⅡ	9
Robert	ⅢⅠ	6

The tally chart shows
the number of pennies
each student in
Mr. Jones's class has.

1. Write a title that explains what the data show.
2. Decide what numbers you will use and write
   them along the bottom of your graph.
3. Draw your bars in the graphs.

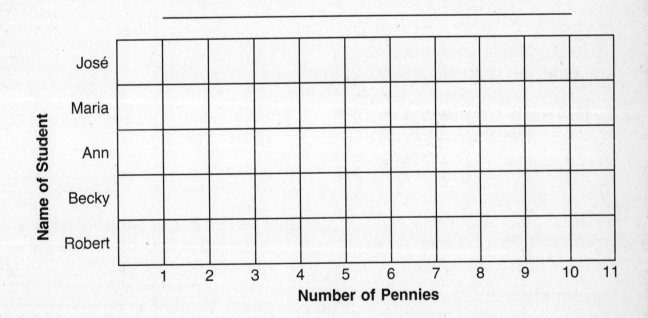

4. Write a summary of your finished graph without using numbers.

_____

_____

Make a Graph

The members of a team collected
aluminum cans for a school project.

Team Member	Cans Collected
Milton	45
Emma	10
Terri	55
Mando	25
Lara	15
Joshua	30

Make a graph to show the data.
Then use the chart and the graph to solve each problem.

1. How many cans did the team
collect in all?

_____

2. How many more cans did Mando
collect than Emma?

_____

3. Milton put 15 cans into each
collecting box he used. How
many boxes did he use?

_____

4. The number of cans Joshua
collected is how many times the
number Lara collected?

_____

5. There were 7 teams. Each had
the same number of members.
How many people in all were
collecting aluminum cans?

_____

6. Terri's brother was on a different
team. She collected 11 times as
many cans as he did. How many
cans did he collect?

_____

NAME

Locating Points on a Grid

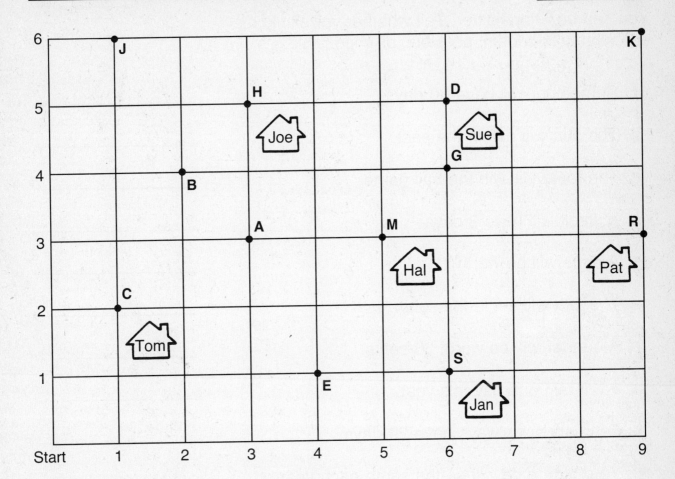

Write the letter for each number pair.

1. (2, 4) _____   **2.** (4, 1) _____   **3.** (9, 6) _____

4. (6, 4) _____   **5.** (3, 3) _____   **6.** (1, 6) _____

Write the number pair for each letter.

7. D _____   **8.** C _____   **9.** R _____

10. H _____   **11.** S _____   **12.** M _____

Solve each problem.

13. Sue lives near (6, 5). Does she live closer to Joe or to Pat?

14. Jan lives near (6, 1). Joe lives near (3, 5). Who lives closer to Hal?

_____   _____

Probabilities

You can be a forecaster. Tell whether you think each event is certain, possible, or impossible.

1. September will have 30 days. _____

2. The sun will rise in the east. _____

3. The team will win the ball game. _____

4. A week will have 8 days. _____

5. A dime will be worth 12 cents. _____

6. The sun will set in the south. _____

7. A quarter will be worth 25 cents. _____

8. It will rain one day this year. _____

9. February will always have 29 days. _____

10. Your birth date will be the same each year. _____

11. A day will have 22 hours. _____

12. One day the temperature will be 55 degrees. _____

13. Someone will be late for school next week. _____

14. May will have 31 days. _____

15. Clocks will have 13 numbers. _____

16. Zero will not be used any more. _____

17. New Year's Day will be the first of January. _____

18. A year will have 12 months. _____

Outcomes

Use the cards to answer the exercises.

If Mario picks a card without looking, are the outcomes equally likely? If not, which is more likely?

1. to pick a triangle or to pick a circle

_____

_____

2. to pick a circle or to pick a square

_____

_____

3. to pick a square or to pick a triangle

_____

_____

4. to pick a star or to pick a triangle

_____

_____

5. to pick a star or to pick a circle

_____

_____

6. to pick a star or to pick a square

_____

_____

SHARPEN
YOUR
SKILLS

Experiments

Answer the following questions. You may
want to do these activities with a partner.

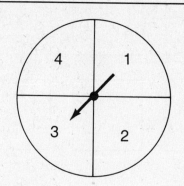

Experiment A

Make a spinner like the one shown.

a. Spin the spinner.

b. Record the number.

1. Is it possible to get a 6? _____

2. Is it possible to get a 4? _____

3. Do you have an equal chance to get a 2 or a 3? _____

4. Are you more likely to get a 4 than a 2? _____

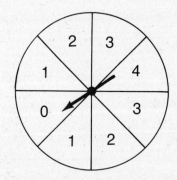

Experiment B

Make a spinner like the one shown.

a. Spin the spinner.

b. Record the number.

5. Is it possible to get a 0? _____

6. Do you have an equal chance to get a 2 or a 4? _____

7. Are you more likely to get a 1, 2, or 3 than a 4 or 0? _____

8. How many times do you think you would get a 2 in 8 spins?

_____

Sharing Money

Divide each amount of money into equal groups.
Use the pictures and draw other pictures to
show your answer. Then write how much is in
each group.

1. $24 into 3 equal groups

_____

2. $36 into 4 equal groups

_____

3. $25 into 5 equal groups

_____

4. $66 into 3 equal groups

$10 $10 $10 $1 $1 $1
$10 $10 $10 $1 $1 $1

_____

5. $32 into 2 equal groups

$10 $10 $10 $1 $1

_____

Problem Solving Solve each problem.

6. Together, 3 girls earned $15 on
Friday and $12 on Saturday. If
they divide the money equally,
how much will each girl receive?

7. The 3 girls want to buy a set of
books for $42. Each girl will pay
the same amount. How much will
each girl pay?

_____ _____

SHARPEN
YOUR
SKILLS

Missing Factors

Match each multiplication sentence with the related division sentence. Then give the missing factor.

1. $8 \times \square = 128$

_____

2. $\square \times 7 = 427$

_____

3. $\square \times 2 = 140$

_____

4. $4 \times \square = 320$

_____

a. $128 \div 8 = 16$
b. $320 \div 4 = 80$
c. $126 \div 3 = 42$
d. $140 \div 2 = 70$
e. $427 \div 7 = 61$
f. $120 \div 2 = 60$
g. $168 \div 4 = 42$
h. $492 \div 6 = 82$
i. $368 \div 8 = 46$
j. $288 \div 9 = 32$
k. $387 \div 9 = 43$
l. $282 \div 6 = 47$

5. $2 \times \square = 120$

_____

6. $\square \times 3 = 126$

_____

7. $9 \times \square = 288$

_____

8. $\square \times 6 = 492$

_____

9. $\square \times 4 = 168$

_____

10. $8 \times \square = 368$

_____

11. $6 \times \square = 282$

_____

12. $\square \times 9 = 387$

_____

Problem Solving · Solve the problem.

13. Jill's class has to spend $39 to rent a bus for a Saturday field trip. If they want to pay just $3 each, how many children must go to pay for the bus?

_____

14. Ronald has $45 to pay for lunches for the trip. How many can he buy if each lunch costs $5? How much more money will he need to buy 15 lunches?

_____

SHARPEN
YOUR
SKILLS

Remainders When Sharing Money

Tell how many are in each group and how many are
left over. Use the pictures and draw other
pictures to show your answer.

1. $45 into 4 equal groups _____

2. $57 into 4 equal groups _____

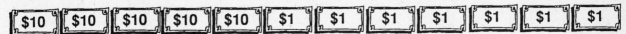

3. $57 into 5 equal groups _____

4. $63 into 5 equal groups _____

5. $27 into 6 equal groups _____

6. $43 into 3 equal groups _____

7. $74 into 6 equal groups _____

Interpret Remainders

Solve each problem.

1. Elena is giving a party. Paper hats are $2 each. How many can she buy if she has $13?

2. Flower pots to decorate the porch cost $4. How many pots could Elena buy for $38? How much money is left over?

3. Elena went to the store with $8. Game books were $3. How many game books could she buy? What is the remainder?

4. For Elena's party, the children went to the zoo. Tickets were $4 each. How many tickets could Elena buy for $25?

5. At the zoo, the guide told Elena's friends that it costs $7 a day to take care of a monkey. How many monkeys can the zoo take care of for $24? How much money is left over?

6. Elena's aunt bought her some new shirts for her birthday. How many shirts would Elena get if the shirts cost $9 each and her aunt had $50 to spend?

7. Elena is saving $5 bills to buy a computer. She had her birthday money changed into $5 bills. How many bills did she have if she received $38 for her birthday?

8. How many more dollars must Elena save before she can get another $5 bill?

Recording Division

Use pictures or play money to divide each
amount of money into equal groups.

1. $64 into 4 equal groups

_____

2. $57 into 5 equal groups

_____

3. $84 into 7 equal groups

_____

4. $58 into 3 equal groups

_____

5. $75 into 6 equal groups

_____

6. $48 into 6 equal groups

_____

7. $83 into 3 equal groups

_____

8. $33 into 4 equal groups

_____

9. $91 into 5 equal groups

_____

10. $42 into 3 equal groups

_____

Mixed Practice   Use paper and pencil or mental math
to find each answer. Write P if you used paper and pencil.
Write M if you used mental math. Then write the answer.

11. $5 \times 8 =$ _____

12. $23 \times 9 =$ _____

13. $81 \div 9 =$ _____

14. $54 \div 6 =$ _____

15. $8 \times 6 =$ _____

16. $27 \div 3 =$ _____

17. $87 \times 6 =$ _____

18. $37 \div 5 =$ _____

19. $91 \div 7 =$ _____

SHARPEN
YOUR
SKILLS

Choose an Operation

Tell whether you should *add, subtract, multiply,*
or *divide.* Then find the answer.

1. The 9 girls at the Chung family
reunion formed a baseball team.
The boys formed a team of 9 and
5 boys did not play. How many
children were at the reunion?

_____

2. Jerry Li and Andy Chan put
sleeping bags on cots. They had
57 bags. When they finished, 8
bags were left. How many bags
had they put on cots?

_____

3. At the reunion, 5 of the families
were named Chung. There were
3 children in each Chung family.
How many of the children were
named Chung?

_____

4. Andy's parents bought meat for a
picnic. They bought 16 chickens
for $3 each. How much money
was left if they paid with a $50
bill?

_____

5. After the picnic, most families
played board games. Four people
could sit at a table. How many
tables were needed for 66 people?

_____

6. Sue and Lin invited 24 friends to a
hayride. Eighteen people came.
How many people did not come?

_____

Critical Thinking Look back at Exercise 5.
If 4 people sat at each table, could all
66 people play board games? Tell why or why not.

7. _____